BEYOND

CONFRONTING GLOBAL COLLAPSE

THE

ENVISIONING A SUSTAINABLE FUTURE

LIMITS

DONELLA H. MEADOWS, DENNIS L. MEADOWS, JØRGEN RANDERS

CHELSEA GREEN PUBLISHING COMPANY
WHITE RIVER JUNCTION, VERMONT

Printed in the United States of America.

Published by Chelsea Green Publishing Company, P.O. Box 428,
White River Junction, Vermont 05001

Beyond the Limits was designed by Kate Mueller,
typeset in New Baskerville, and printed by
Royal Book Manufacturing, Inc.

10 9 8 7 6 5 4 3

Library of Congress Cataloging-in-Publication Data

Meadows, Donella H.

Beyond the limits: confronting global collapse, envisioning a sustainable future / Donella H. Meadows, Dennis L. Meadows, Jørgen Randers

p. cm.

ISBN 0-930031-55-5 (cl) — ISBN 0-930031-62-8 (pbk)

1. Economic development—Economic aspects. 2. Population—Economic aspects. 3. Pollution—Economic aspects. I. Meadows, Dennis L. II. Randers, Jørgen. III. Title.

HD75.6M43 1992

338.9dc20 91-46920

CIP

ACKNOWLEDGMENTS

Inspiration

This book is dedicated to:

- Aurelio Peccei, founder of The Club of Rome. His profound concern for the world and his faith in humanity inspired us and many others to care about and address the prospects for the long-term future.
- Jay W. Forrester, Professor Emeritus of the Sloan School of Management at MIT and our teacher. He designed the prototype of the computer model we have used in this book, and his systems insights have helped us to understand the behaviors of economic and environmental systems.

Administration

Critical to the preparation of this book were the following people, to whom we extend our heartfelt gratitude:

- Ian and Margo Baldwin of Chelsea Green Publishing Company for taking this project on and for devoting their own energies and the resources of their company so unstintingly toward its completion;
- Angele Cook of the University of New Hampshire and Anita Brown and Mardi McGregor of Dartmouth College for providing constant and cheerful logistical support;
- Lew Feldstein and the New Hampshire Charitable Trust, for supporting creation of a policy research center in New Hampshire.
- Suzanne MacDonald for hosting extended writing sessions at her house and for sustaining and encouraging exhausted writers;
- Peter Matson of Sterling Lord Literistic for unsnarling this work from the old *Limits* and moving it forward to publication;
- Marie and Engelke Randers for loaning their husband and father for many weeks to a project across the ocean;
- The residents of Foundation Farm, for keeping the home fires burning while one of their farmers was busy with a book;
- James Hornig for creating an environment at Dartmouth College that gave us the material and intellectual support necessary to research and prepare this book;
- Barry Richmond and Steve Peterson of High Performance Systems, Inc. for the STELLA II© software that has made the World3 model so much more accessible than it was twenty years ago.
- Readers and commenters including: William W. Behrens III, Allen Boorstein, Hartmut Bossel, Lester Brown, Chester Cooper, Herman Daly, Joan Davis, Judy Gabriel, Jay Harris, John Harte, James Hornig, Nathan Keyfitz, Niels Meyer, Don Michael, Mario Molina, Russell Peterson, Aromar Revi, John Sterman, and Steve Viederman. We have not been able to please them all in every detail, but their comments were frank, thought-provoking, and helpful. Their quick and energetic responses despite very busy schedules testify to their commitment to resolving the issues raised in this book.

Donation

Funding for the preparation of *Beyond the Limits* was provided by the

Pew Scholars Program, Jane and Allen Boorstein, Jay Harris, and William Welsh.

Perspiration

The team that did the research, ran the computer model, created the graphics, and wrote this book consisted of:

- Dr. Bert de Vries, National Institute of Public Health and Environmental Protection (RIVM), the Netherlands;
- Thomas Fiddaman, Institute for Policy and Social Science Research, University of New Hampshire, USA;
- Dr. Dennis Meadows, Institute for Policy and Social Science Research, University of New Hampshire, USA;
- Dr. Donella H. Meadows, Environmental Studies Program, Dartmouth College, USA;
- Dr. Jørgen Randers, Chairman, S. Sejersted Bodtker & Co. AS, Norway;
- Diana Wright, Environmental Studies Program, Dartmouth College, USA.

The original project that produced the World3 computer model and *The Limits to Growth* took place in the System Dynamics Group of the Sloan School of Management at the Massachusetts Institute of Technology. It was commissioned by The Club of Rome and funded by the Volkswagen Foundation. The team consisted of:

Dr. Alison A. Anderson (USA)
Dr. Jay M. Anderson (USA)
Ilyas Bayar (Turkey)
Dr. William W. Behrens III (USA)
Farhad Hakimzadeh (Iran)
Dr. Steffen Harbordt (Germany)
Judith A. Machen (USA)
Dr. Dennis L. Meadows (USA)
Dr. Donella H. Meadows (USA)
Dr. Peter Milling (Germany)
Nirmala S. Murthy (India)
Dr. Roger F. Naill (USA)
Dr. Jørgen Randers (Norway)
Stephen Schantzis (USA)
Dr. John A. Seeger (USA)
Marilyn Williams (USA)
Dr. Erich K. O. Zahn (Germany)

CONTENTS

FOREWORD

We all can learn some lessons from this book, especially we economists. We can learn about the background against which economic processes are developing and about the space in which they take place, our planet Earth. That background, that space, is large in comparison to the problems economists usually deal with, but it is finite, and everything economic has to be done on, in, or around it.

Two things are unlimited: the number of generations we should feel responsible for and our inventiveness. The first provides us with a challenge: to feed and provide for not only the present but all future generations from the earth's finite flow of natural resources. The second, our inventiveness, may create ideas and policies that will contribute to meeting that challenge.

Our responsibility to all generations extends especially to those now living on poor continents or in the poorest quarters of cities on all continents. In the present and the future it extends to more than ensuring food and material provisions; it also extends to keeping the environment clean.

The time is past when incomes are becoming unequal globally. But at the present rate it would still take far too long for them to become equal: five centuries. Whether the highest present incomes can be maintained is very doubtful. Market economies are obviously in need of some intervention in order to provide public goods, to avoid too much inequality, and to approach sustainability.

It is the great merit of *Beyond the Limits* that it shows us where and when we may reach the frontiers of the possible and thus clarifies the conditions under which sustainable development, a clean environment, and equitable incomes can be organized. It shows that there are exciting possibilities, and that they are limited, more so than some economists think. It reveals that the possible average sustainable income level is lower today than twenty years ago. That is the consequence of our failure to understand the limits to the use of natural resources. And the book also shows us where human creativity has improved our prospects, as in energy efficiency, resource recycling, and increases in the average length of human life.

As economists we must be grateful to these authors for showing us where the present path of human development threatens to exceed the limits, and for illustrating the contributions economics and other disciplines must make to meet the great human challenge of avoiding war, famine, disease, and pollution, and of building a sustainable future.

Jan Tinbergen
Nobel Laureate, Economics

PREFACE

Twenty years ago we wrote a book called *The Limits to Growth*.[1] It described the prospects for growth in the human population and the global economy during the coming century. In it we raised questions such as: What will happen if growth in the world's population continues unchecked? What will be the environmental consequences if economic growth continues at its current pace? What can be done to ensure a human economy that provides sufficiently for all and that also fits within the physical limits of the Earth?

We had been commissioned to examine these questions by The Club of Rome, an international group of distinguished businessmen, statesmen, and scientists. They asked us to undertake a two-year study at the Massachusetts Institute of Technology to investigate the long-term causes and consequences of growth in population, industrial capital, food production, resource consumption, and pollution. To keep track of these interacting entities and to project their possible paths into the future we created a computer model called World3.[2]

The results of our study were described for the general public[3] in

The Limits to Growth. That book created a furor. The combination of the computer, MIT, and The Club of Rome pronouncing upon humanity's future had an irresistible dramatic appeal. Newspaper headlines announced:

> A COMPUTER LOOKS AHEAD AND SHUDDERS
> STUDY SEES DISASTER BY YEAR 2100
> SCIENTISTS WARN OF GLOBAL CATASTROPHE.[4]

Our book was debated by parliaments and scientific societies. One major oil company sponsored a series of advertisements criticizing it; another set up an annual prize for the best studies expanding upon it. *The Limits to Growth* inspired some high praise, many thoughtful reviews, and a flurry of attacks from the left, the right, and the middle of mainstream economics.

The book was interpreted by many as a prediction of doom, but it was not a prediction at all. It was not about a preordained future. It was about a choice. It contained a warning, to be sure, but also a message of promise. Here are the three summary conclusions we wrote in 1972. The second of them is the promise, a very optimistic one, but our analysis justified it then and still justifies it now. Perhaps we should have listed it first.

1. If the present growth trends in world population, industrialization, pollution, food production, and resource depletion continue unchanged, the limits to growth on this planet will be reached sometime within the next 100 years. The most probable result will be a sudden and uncontrollable decline in both population and industrial capacity.

2. It is possible to alter these growth trends and to establish a condition of ecological and economic stability that is sustainable far into the future. The state of global equilibrium could be designed so that the basic material needs of each person on earth are satisfied and each person has an equal opportunity to realize his or her individual human potential.

3. If the world's people decide to strive for this second outcome rather than the first, the sooner they begin working to attain it, the greater will be their chances of success.[5]

To us those conclusions spelled out not doom but challenge–how to bring about a society that is materially sufficient, socially equitable, and ecologically sustainable, and one that is more satisfying in human terms than the growth-obsessed society of today.

In one way and another, we've been working on that challenge ever since. Millions of other people have been working on it too. They've been exploring energy efficiency and new materials, nonviolent conflict resolution and grassroots community development, pollution prevention in factories and recycling in towns, ecological agriculture and international protocols to protect the ozone layer. Much has happened in twenty years to bring about technologies, concepts, and institutions that can create a sustainable future. And much has happened to perpetuate the desperate poverty, the waste of resources, the accumulation of toxins, and the destruction of nature that are tearing down the support capacity of the earth.

When we began working on the present book, we simply intended to document those countervailing trends in order to update *The Limits to Growth* for its reissue on its twentieth anniversary. We soon discovered that we had to do more than that. As we compiled the numbers, reran the computer model, and reflected on what we had learned over two decades, we realized that the passage of time and the continuation of many growth trends had brought the human society to a new position relative to its limits.

In 1971 we concluded that the physical limits to human use of materials and energy were somewhere decades ahead. In 1991, when we looked again at the data, the computer model, and our own experience of the world, we realized that in spite of the world's improved technologies, the greater awareness, the stronger environment policies, many resource and pollution flows had grown beyond their sustainable limits.

That conclusion came as a surprise to us, and yet not really a surprise. In a way we had known it all along. We had seen for ourselves the leveled forests, the gullies in the croplands, the rivers brown with silt. We knew the chemistry of the ozone layer and the greenhouse effect. The media had chronicled the statistics of global fisheries, groundwater drawdowns, and the extinction of species. We discovered, as we began to talk to colleagues about the world being "beyond the limits," that they did not question that conclusion. We found many places in the literature of the past twenty years where authors had suggested that resource and pollution flows had grown too far, some of which we have quoted in this book.

But until we started updating *The Limits to Growth* we had not let our minds fully absorb the message. The human world is beyond its limits. The present way of doing things is unsustainable. The future, to be viable at all, must be one of drawing back, easing down, healing. Poverty cannot be ended by indefinite material growth; it will have to be addressed while the material human economy contracts. Like everyone else, we didn't really want to come to these conclusions.

But the more we compiled the numbers, the more they gave us that message, loud and clear. With some trepidation we turned to World3, the computer model that had helped us twenty years before to integrate the global data and to work through their long-term implications. We were afraid that we would no longer be able to find in the model any possibility of a believable, sufficient, sustainable future for all the world's people.

But, as it turned out, we could. World3 showed us that in twenty years some options for sustainability have narrowed, but others have opened up. Given some of the technologies and institutions invented over those twenty years, there are real possibilities for reducing the streams of resources consumed and pollutants generated by the human economy while increasing the quality of human life. It is even possible, we concluded, to eliminate poverty while accommodating the population growth already implicit in present population age structures—but not if population growth goes on indefinitely, not if it goes on for long, and not without rapid improvements in the efficiency of material and energy use and in the equity of material and energy distribution.

As far as we can tell from the global data, from the World3 model, and from all we have learned in the past twenty years, the three conclusions we drew in *The Limits to Growth* are still valid, but they need to be strengthened. Now we would write them this way:

1. Human use of many essential resources and generation of many kinds of pollutants have already surpassed rates that are physically

sustainable. Without significant reductions in material and energy flows, there will be in the coming decades an uncontrolled decline in per capita food output, energy use, and industrial production.

2. This decline is not inevitable. To avoid it two changes are necessary. The first is a comprehensive revision of policies and practices that perpetuate growth in material consumption and in population. The second is a rapid, drastic increase in the efficiency with which materials and energy are used.

3. A sustainable society is still technically and economically possible. It could be much more desirable than a society that tries to solve its problems by constant expansion. The transition to a sustainable society requires a careful balance between long-term and short-term goals and an emphasis on sufficiency, equity, and quality of life rather than on quantity of output. It requires more than productivity and more than technology; it also requires maturity, compassion, and wisdom.

These conclusions constitute a conditional warning, not a dire prediction. They offer a living choice, not a death sentence. The choice isn't necessarily a gloomy one. It does not mean that the poor must be frozen in their poverty or that the rich must become poor. It could actually mean achieving at last the goals that humanity has been pursuing in continuous attempts to maintain physical growth.

We hope the world will make a choice for sustainability. That is why we are writing this book. But we do not minimize the gravity or the difficulty of that choice. We think a transition to a sustainable world is technically and economically possible, maybe even easy, but we also know it is psychologically and politically daunting. So much hope, so many personal identities, so much of modern industrial culture has been built upon the premise of perpetual material growth.

A perceptive teacher, watching his students react to the idea that there are limits, once wrote:

> When most of us are presented with the ultimata of potential disaster, when we hear that we "must" choose some form of planned stability, when we face the "necessity" of a designed sustainable

> state, we are being bereaved, whether or not we fully realize it. When cast upon our own resources in this way we feel, we intuit, a kind of cosmic loneliness that we could not have foreseen. We become orphans. We no longer see ourselves as children of a cosmic order or the beneficiaries of the historical process. Limits to growth denies all that. It tell us, perhaps for the first time in our experience, that the only plan must be our own. With one stroke it strips us of the assurance offered by past forms of Providence and progress and with another it thrusts into our reluctant hands the responsibility for the future.[6]

We went through that entire emotional sequence–grief, loneliness, reluctant responsibility–when we worked on The Club of Rome project twenty years ago. Many other people, through many other kinds of formative events, have gone through a similar sequence. It can be survived. It can even open up new horizons and suggest exciting futures. Those futures will never come to be, however, until the world as a whole turns to face them. The ideas of limits, sustainability, sufficiency, equity, and efficiency are not barriers, not obstacles, not threats. They are guides to a new world. Sustainability, not better weapons or struggles for power or material accumulation, is the ultimate challenge to the energy and creativity of the human race.

We think the human race is up to the challenge. We think that a better world is possible, and that the acceptance of physical limits is the first step toward getting there. We see "easing down" from unsustainability not as a sacrifice, but as an opportunity to stop battering against the earth's limits and to start transcending self-imposed and unnecessary limits in human institutions, mindsets, beliefs, and ethics. That is why we finally decided not just to update and reissue *The Limits to Growth*, but to rewrite it completely and to call it *Beyond the Limits*.

Donella H. Meadows
Dennis L. Meadows
Jørgen Randers

Durham, New Hampshire
November 1991

A Note on Language

In this book we use *billion* in the American sense to mean *1000 million*, equivalent to the European *milliard*.

We distinguish U.S. *tons* (2000 lb or 907 kg) from European *metric tons* or *tonnes* (2205 lb or 1000 kg).

Capital always means here physical plant: the hardware, machines, factories, and equipment that produce economic goods and services. If we refer to the money needed to finance construction of physical plant, we call it *financial capital.*

We use a number of terms from the field of systems analysis throughout this book. We define each one at first use, and we have summarized all of them in a glossary at the end. Examples of these terms are: *system, structure, overshoot, exponential growth, feedback loop, source, sink,* and *throughput.*

Like everyone, we have trouble with the choice of words to designate different regions of the world. We object to the words *developed* and *developing* for reasons that will become evident as we make a case here for new and different development patterns. The terms *First, Second,* and *Third Worlds* distinguish between the Western market economies, the former centrally planned economies of Europe, and the "rest of the world," but that distinction is Western-biased and rapidly waning in relevance. *North* and *South* are geographically inaccurate but value-free designations often used in United Nations documents to refer, loosely, to richer and poorer regions. Since we quote here from many sources in different contexts, we will use all the above terms from time to time.

But the distinction we think is most accurate for our purposes is between cultures that are *industrialized* and *less-industrialized.* We mean to signify by those terms the degree to which different parts of the world (including whole nations and also subsets of populations within nations) have undergone the Industrial Revolution: the degree to which their economies have shifted from agriculture-dominance to industry- and service-dominance, the degree to which their main energy sources are fossil and nuclear fuels, the degree to which they have absorbed the labor patterns, family sizes, consumption habits, and mindset of the modern technological culture.

Finally, the most important distinction we shall make in this book is the one between *growth* and *development.*

> Following the dictionary distinction . . . TO GROW means to increase in size by the assimilation or accretion of materials. TO DEVELOP means to expand or realize the potentialities of; to bring to a fuller, greater, or better state. When something grows it gets quantitatively bigger; when it develops it gets qualitatively better, or at least different. Quantitative growth and qualitative improvement follow different laws. Our planet develops over time without growing. Our economy, a subsystem of the finite and non-growing earth, must eventually adapt to a similar pattern of development.[1]

We think there is no more important distinction to keep straight than that one. It tells us that, although there are limits to growth, there need be no limits to development.

chapter 1

OVERSHOOT

The future is no longer what it was thought to be, or what it might have been if humans had known how to use their brains and their opportunities more effectively. But the future can still become what we reasonably and realistically want.

Aurelio Peccei[1]

To overshoot means to go beyond limits inadvertently, without meaning to do so. Daily life is full of small and not-so-small overshoots. A car on an icy road can slide past a stop sign. If you eat or drink too fast, you can go too far before your body sends unmistakable signals that you should stop.

On a larger scale a fishing fleet can become so large and efficient that it depletes the fish population upon which it depends. Developers can put up more condominiums than people are able or willing to buy. An electric utility can build more generating capacity than the economy can use.

The underlying causes of overshoot are always the same. First there is rapid motion, action, or change. Second there is some sort of limit or barrier, beyond which the motion or action or change should not go. Third there is a difficulty in control, because of inattention, faulty data, delayed feedback, inadequate information, slow response, or simple momentum. The driver goes too fast for the brakes to work in time on the slippery road. The fishing fleet builds up its capacity faster than it gets reliable information about the state of the fish population. The utility decides too rapidly under conditions of too much uncertainty to start construction projects that take too long to complete.

This book is about overshoot on a much larger scale, namely the scale at which the human population and economy extract resources from the earth and emit pollution and wastes to the environment. Many of these rates of extraction and emission have grown to be unsupportable. The environment cannot sustain them. Human society has overshot its limits, for the same reasons that other overshoots occur. Changes are too fast. Signals are late, incomplete, distorted, ignored, or denied. Momentum is great. Responses are slow.

After overshoot can come a number of possible consequences. One of them, of course, is some kind of crash. Another is a deliberate turnaround, a correction, a careful easing down. This book explores these two possibilities as they apply to the human society and the supporting planet. We believe that a correction is possible and that it could lead to a desirable, sufficient, equitable, and sustainable future. We also believe that if a correction is not made, a collapse of some sort is not only possible but certain, and that it could occur within the lifetimes of many who are alive today.

Those are enormous claims. How did we arrive at them?

We looked at the long-term implications of the present rates of change in the human society with four kinds of viewing devices—four different lenses to help us focus on the world in different ways, just as the lenses of a microscope and a telescope enable one to see different things. Three of these viewing devices are fairly common and easy to describe and to hand on to others, namely: standard scientific and economic theory about the global system; statistical information on the world's resources and environment; and a computer model to help us integrate that information. Much of this book describes those lenses, how we used them, and what they allowed us to see.

Our fourth lens, probably the most important one, was our "worldview," or paradigm, or fundamental way of looking. Everybody has a worldview. It is always the most important determinant of what one sees. And it is almost impossible to describe. Ours was formed by the

Western industrial societies in which we grew up, by our scientific and economic training, and by the considerable education we have received from colleagues in resource management with whom we have worked in many parts of the world. But the most important part of our way of looking, the part that is perhaps least widely shared, is our systems viewpoint.

A systems viewpoint is not necessarily a better one than any other, just a different one. Like any viewpoint, like the top of any hill you climb, it lets you see some things you would never have noticed from any other place, and it blocks the view of other things. Systems training has taught us to see the world as a set of unfolding dynamic behavior patterns, such as growth, decline, oscillation, overshoot. It has taught us to focus on interconnections. We see the economy and the environment as *one system*. We see stocks and flows and feedbacks and thresholds in that system, all of which influence the way the system behaves.

The systems viewpoint is by no means the only useful way to see the world, and it is not the only one we use. But it's one we find particularly informative and exciting. It lets us approach problems in new ways and discover unsuspected options. We intend to share its principal concepts with you here, so you can see what we see through that lens and form your own conclusions about the state of the world and the choices for the future.

The structure of this book follows the logic of our analysis of the global system. You do not need high mathematics to understand it, and you don't need to be a computer expert. We have already said that overshoot comes from the combination of rapid change, limits or barriers to that change, and imperfections in the signals about or the responses to those limits. We will look at the global situation in that order—first at global change, then at planetary limits, then at the ways human society learns about and responds to the limits.

We start in the next chapter with the rapid change, which in the global system comes most basically from population and economic growth. Growth has been the dominant behavior of the socioeconomic system for more than two hundred years. For example, Figure 1-1 shows the growth of world population, which is still surging upward faster and faster, despite recent drops in birth rates in some countries.

Figure 1-1 WORLD POPULATION

Billions of people

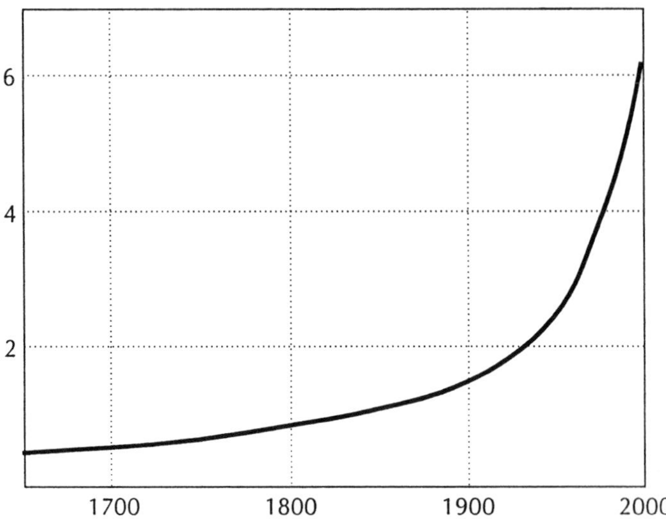

World population has been growing exponentially since the beginning of the Industrial Revolution. In 1991 the world population growth rate was estimated to be 1.7%, corresponding to a doubling time of 40 years. (*Sources: United Nations; D. J. Bogue.*)

Industrial production is growing too, as illustrated in Figure 1-2, even faster than population, in spite of some dips at times of major oil price rises. Industrial production has risen slightly faster than population, resulting in a slow and bumpy increase in the average material standard of living of the population.

Many kinds of pollution are also growing. Figure 1-3 shows just one, the rise of carbon dioxide in the earth's atmosphere, a result of human fossil fuel burning and forest clearing.

Other graphs throughout this book illustrate growth in fertilizer use, cities, energy consumption, materials use, and many other physical manifestations of human activity on the planet. Not everything is growing at the same rate. The rate of increase of oil consumption worldwide has slowed, for instance, while the rate of increase of natural gas consumption has accelerated. Just a few of the material changes of the past

Figure 1-2 World Industrial Production

Index (1963 = 100)

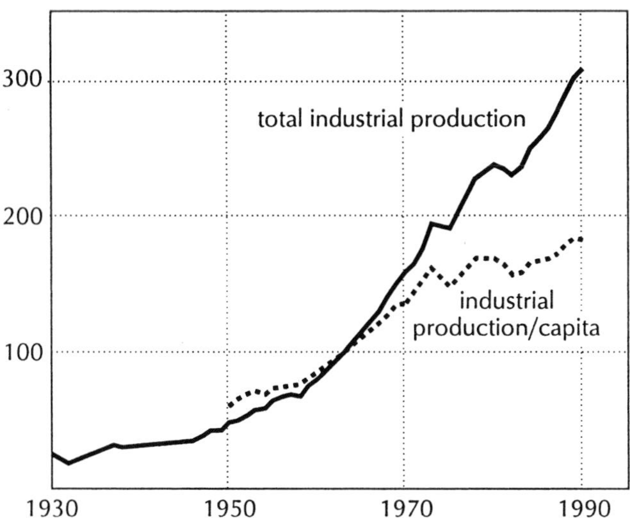

World industrial production, relative to the base year 1963, also shows clear exponential increase, despite fluctuations due to oil price shocks. The 1970–1990 growth rate in total production has averaged 3.3% per year. The per capita growth rate has been 1.5% per year. (*Sources*: *United Nations; Population Reference Bureau.*)

twenty years are shown in Table 1-1. As you can see, the amount of growth varies, but growth continues to be a dominant pattern.

The predominance of growth in human activity comes as no surprise. In fact most people see it as something to celebrate. Most societies, rich or poor, seek some kind of expansion as a remedy for their most immediate and important problems. In the rich world economic growth is believed to be necessary for employment, social mobility, and technical advance. In the poor world economic growth seems the only way out of poverty. And a poor family sees that many children can be a source not only of joy, but also of hope for economic security. Until other solutions are found for the legitimate problems of the world, people will cling to the idea that growth is the key to a better future, and they will do all they can to produce more growth.

Figure 1-3 CARBON DIOXIDE CONCENTRATION IN THE ATMOSPHERE

Parts per million by volume

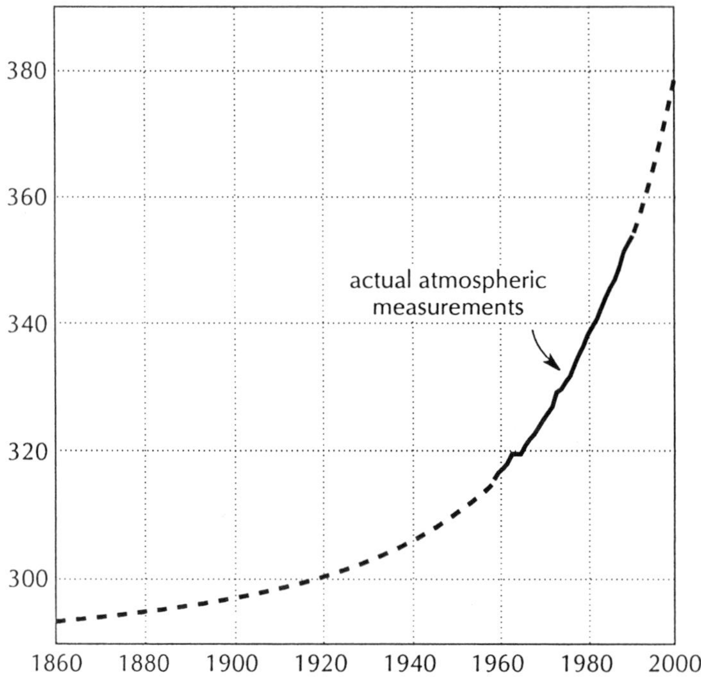

The concentration of carbon dioxide in the atmosphere has risen from roughly 290 parts per million in the last century to over 350 parts per million, and it continues on its exponential growth path. The sources of the carbon dioxide buildup are human fossil fuel burning and forest destruction. The possible consequence is global climate change. (*Sources: L. Machta; T. A. Boden.*)

Those are the psychological and institutional reasons for growth. There are also *structural* reasons, built into the very connections that hold the population and economy together. In Chapter 2 we discuss these structural causes of growth, their implications, why growth is such a dominant behavior of the world system, and why growth is solving only inefficiently, if at all, the problems it is expected to address.

Growth can solve some problems, but it creates others. That is because of limits, the subject of Chapter 3.

Table 1-1 Worldwide Growth in Selected Human Activities and Products 1970–1990

	1970	*1990*
Human population	3.6 billion	5.3 billion
Registered automobiles	250 million	560 million
Kilometers driven/year (OECD countries only)		
by passenger cars	2584 billion	4489 billion
by trucks	666 billion	1536 billion
Oil consumption/year	17 billion barrels	24 billion barrels
Natural gas consumption/year	31 trillion cubic feet	70 trillion cubic feet
Coal consumption/year	2.3 billion tons	5.2 billion tons
Electric generating capacity	1.1 billion kilowatts	2.6 billion kilowatts
Electricity generation/year by nuclear power plants	79 terawatt-hours	1884 terawatt-hours
Soft drink consumption/year (U.S. only)	150 million barrels	364 million barrels
Beer consumption/year (U.S. only)	125 million barrels	187 million barrels
Aluminum used/year for beer and soft drink containers (U.S. only)	72,700 tonnes	1,251,900 tonnes
Municipal waste generated/year (OECD countries only)	302 million tonnes	420 million tonnes

The earth is finite. Growth of anything physical, including the human population and its cars and buildings and smokestacks, cannot continue forever. But the important limits to growth are not limits to population, cars, buildings, or smokestacks, at least not directly. They are limits to *throughput*—to the flows of energy and materials needed to keep people, cars, buildings, and smokestacks functioning.

The human population and economy depend upon constant flows of air, water, food, raw materials, and fossil fuels from the earth. They constantly emit wastes and pollution back to the earth. The limits to growth are limits to the ability of the planetary *sources* to provide those streams of materials and energy, and limits to the ability of the planetary *sinks* to absorb the pollution and waste.

In Chapter 3 we examine, through the global statistical data, the condition of the earth's sources and sinks. The evidence in the chapter makes two points, which form a classic bad news/good news combination.

The bad news is that many crucial sources are declining and degrading and many sinks are overflowing. The throughput flows that maintain the human economy cannot be maintained at their current rates indefinitely, or even for very much longer. The good news is that the current high rates of throughput are not necessary to support a decent standard of living for all the world's people. Technical changes and efficiencies are possible and available, which can help maintain production of final goods and services while reducing greatly the burden on the planet. There are many choices, many ways of bringing the human society back from beyond its throughput limits.

But that is not the end of the story. Those choices are not being made, at least not strongly enough to make a difference soon enough. They are not being made because there is no obvious or immediate reason to make them. That is the subject of Chapter 4, which looks at the signals that warn human society of its condition of overshoot and the speed with which society can respond.

In Chapter 4 we turn to the computer model, World3. We describe the purpose, structure, and behavior of World3. We show what happens when we use the model to simulate the world system as it might evolve if there were no structural changes, no extraordinary efforts to see ahead, to improve signals, or to solve problems before they become critical. The result of those simulations is not only overshoot, but collapse.

Fortunately there is evidence that the real human world is more competent than the simplified model world of Chapter 4. In Chapter 5 we tell the best story we know about humanity's ability to look ahead, sense a limit, and pull back. We describe the world's actual response to the news of a deteriorating stratospheric ozone layer. The story is important, we think, for several reasons. First and most important, it offers hope. Second, it illustrates every structural point we have made about the global system: rapid growth, limits, slow responses (in both the political system and the natural system), and overshoot. Third, the conclusion of the story is not yet clear and won't be for decades, and so it becomes a cautionary tale, an illustration of how tricky it is to guide the complex human economy within the even more complex systems of the planet with imperfect understanding, lack of foresight, and high momentum.

In Chapter 6 we return to the World3 model and begin to build into it various hypotheses about human cleverness. We concentrate in that chapter on the forms of cleverness in which many people have the greatest faith—technology and markets. Important features of those two remarkable human response capacities are already contained within World3, but in Chapter 6 we strengthen them. We ask: What would happen if the world society began to allocate its resources seriously to the technologies of pollution control, land preservation, human health, materials recycling, and resource-use efficiency?

We discover that these measures help considerably. But they are not enough. They fall short because technology-market responses are themselves delayed and imperfect. They take time, they take capital, they themselves require material and energy flows to sustain them, and they can be overwhelmed by the ever-increasing changes induced by growth. Technological progress and market flexibility will be necessary, we believe, to bring the world to sustainability. But something more is required. That is the subject of Chapter 7.

In Chapter 7 we use World3 to see what happens if human beings supplement their cleverness with wisdom. We assume two definitions of "enough," one having to do with material consumption, the other related to desired family size. With these changes, combined with the technical changes we assumed in Chapter 6, the model world population stabilizes at about 8 billion. All those 8 billion people achieve a level of material welfare roughly equivalent to that of present-day Europe. And, given reasonable assumptions about future market efficiency and technical advance, the material and energy throughputs needed by that model world can be maintained by the planet indefinitely. In this simulation, overshoot is transformed into sustainability.

Sustainability is a concept so foreign to the present growth-acclimated world that we take some time in Chapter 7 to define it and to outline what a world of sustainability might be like—and what it might *not* be like. We see no reason why a sustainable world would or could leave anyone living in poverty. Quite the contrary, we think such a world would have both the opportunity and the necessity to provide material security to all its people at higher standards than they have today. We don't think a sustainable society need be stagnant, boring, fixed, or unadaptive. It need not be rigidly or centrally controlled, or uniform, or undiverse, or undemocratic. What it could be is a world that would have the time and resources to correct its mistakes, to innovate, and to develop without growing beyond its limits.

The concluding chapter derives more from our mental models than from the data or computer model; it is our personal attempt to envision a sustainable state and to imagine how to get there from here. We know that will be a complex task. In fact we think it will be a revolution as profound as the Agricultural and Industrial Revolutions. We appreciate the difficulty of finding sustainable solutions to problems like poverty and employment, for which growth has been, so far, the world's only hope. But we also know that growth is not doing an effective job of solving those problems, that growth is in any case unsustainable, and that other solutions can be found.

Everything we have learned from the global data, from the computer, and from our own training and experience, tells us that the possible paths into the future have narrowed in the past twenty years as

Figure 1-4 ALTERNATIVE PROJECTIONS FOR GLOBAL POPULATION AND CONSUMER GOODS PER CAPITA THROUGH 2100

This figure superimposes all the World3 scenarios shown in this book to illustrate the wide range of possible paths for two important variables—population and per capita consumer goods. Some scenarios show decline; others characterize a society that has achieved a stable population with a high and sustainable standard of living.

FIGURE 1-4

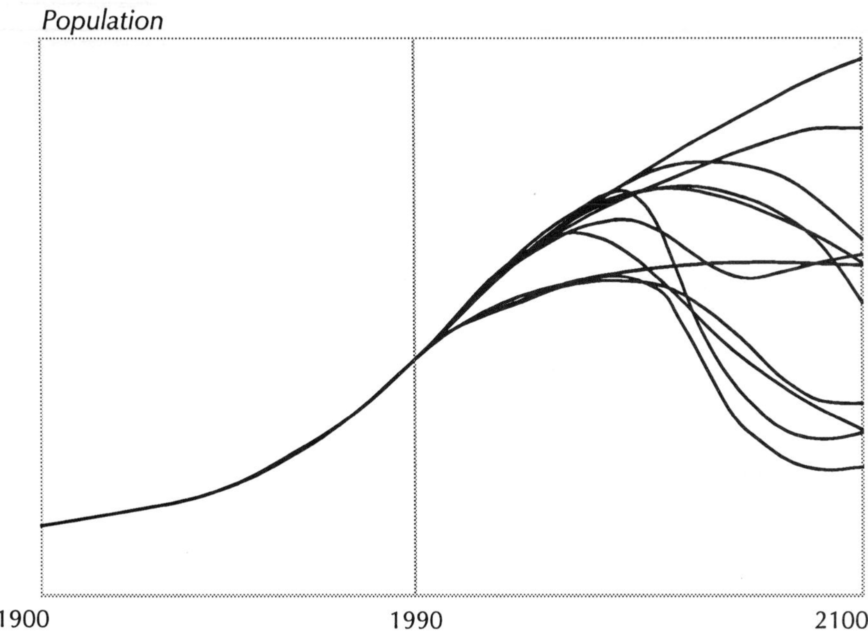
Population
1900
1990
2100

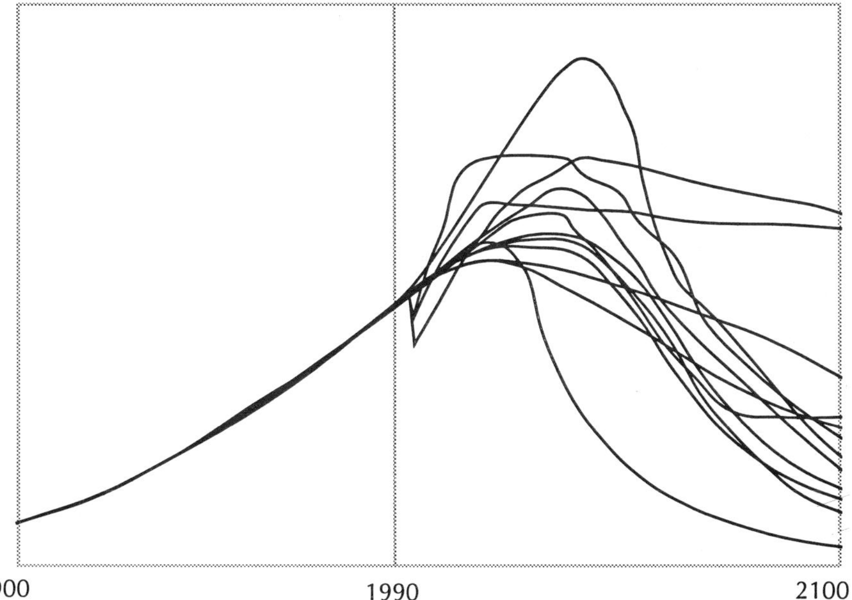
Material standard of living
1900
1990
2100

human society has grown beyond its limits. But there are still many choices, and those choices are crucial. Figure 1-4 illustrates the enormous range of possibilities. The figure was derived by superimposing the curves for human population and for consumer goods per capita generated by all the computer scenarios we present later in this book.

It shows a great variety of future paths. They include various kinds of collapse, and also smooth transitions to more or less sustainable states. They do not include continuous growth. The choices are to bring the burden of human activities upon the earth down to a sustainable level through human choice, human technology, and human organization, or to let nature force the reduction through lack of food, energy, or materials, or an increasingly unsound environment.

Twenty years ago when we wrote *The Limits to Growth* we started with a quotation from U Thant, who was then Secretary-General of the United Nations:

> I do not wish to seem overdramatic, but I can only conclude from the information that is available to me as Secretary-General, that the Members of the United Nations have perhaps ten years left in which to subordinate their ancient quarrels and launch a global partnership to curb the arms race, to improve the human environment, to defuse the population explosion, and to supply the required momentum to development efforts. If such a global partnership is not forged within the next decade, then I very much fear that the problems I have mentioned will have reached such staggering proportions that they will be beyond our capacity to control.

Perhaps, we thought, as we prepared this twentieth-year sequel, we should substitute a more recent and positive vision of the future, such as this one from the World Commission on Environment and Development:

> Humanity has the ability to make development sustainable—to ensure that it meets the needs of the present without compromising the ability of future generations to meet their own needs.[2]

But then we thought again. Perhaps U Thant was right. Perhaps he was premature, and the time that makes him right has now arrived. Or perhaps the best summary of the present situation is the juxtaposition of both these quotes.

Together they capture, according to our analysis and judgment, the enormous range of possible futures and the importance of the choices that are still to be made.

chapter 2

The Driving Force: Exponential Growth

I find to my personal horror that I have not been immune to naivete about exponential functions. . . . While I have been aware that the interlinked problems of loss of biological diversity, tropical deforestation, forest dieback in the Northern Hemisphere and climate change are growing exponentially, it is only this very year that I think I have truly internalized how rapid their accelerating threat really is.

Thomas E. Lovejoy[1]

The first cause of overshoot is rapid motion, growth, or change. In the global system population, food production, industrial production, consumption of resources, and pollution are all growing. Furthermore they are growing more and more rapidly. Their increase follows a pattern that mathematicians call *exponential growth.*

Many human activities, from use of fertilizer to expansion of cities, can be approximated by exponential growth curves (see Figures 2-1 and 2-2). The curves may be interrupted by weather or economic fluctuations or technical change or civil disruption, but on the whole exponential growth has been a prominent and usually welcome pattern of human activity since the industrial revolution.

Exponential growth is the driving force causing the human economy to approach the physical limits of the earth. It is culturally ingrained and structurally inherent in the global system, and the causal structure that produces it is at the core of the World3 model. Therefore we need to begin with an understanding of its mathematics, its causes, and its way of unfolding over time.

Figure 2-1 WORLD FERTILIZER CONSUMPTION

Million metric tons per year

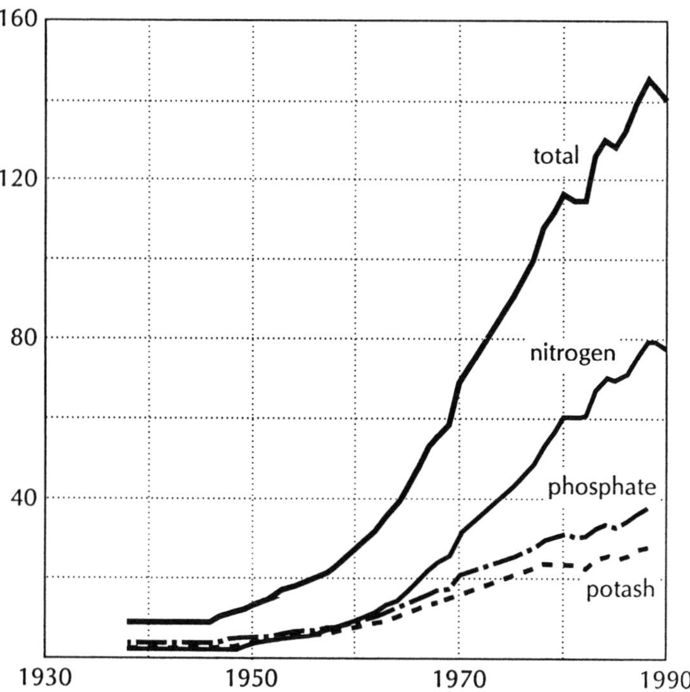

World fertilizer consumption is increasing exponentially with a doubling time of about 10 years before 1970, and of about 15 years after 1970. Total use is now 15 times greater than it was at the end of World War II. (*Source: United Nations.*)

The Mathematics of Exponential Growth

Take a piece of paper and fold it in half. You've just doubled its thickness. Fold it in half again to make it 4 times its original thickness. Assuming you could go on folding the paper like that for a total of 40 times, how thick do you think it would get to be? Less than a foot? Between a foot and 10 feet? Between 10 feet and a mile?

In fact you could not fold a paper 40 times, but if somehow its thickness could be doubled 40 times over, it would make a pile of paper high enough to reach from the earth to the moon.[2]

Figure 2-2 World Urban Population

Billions of people

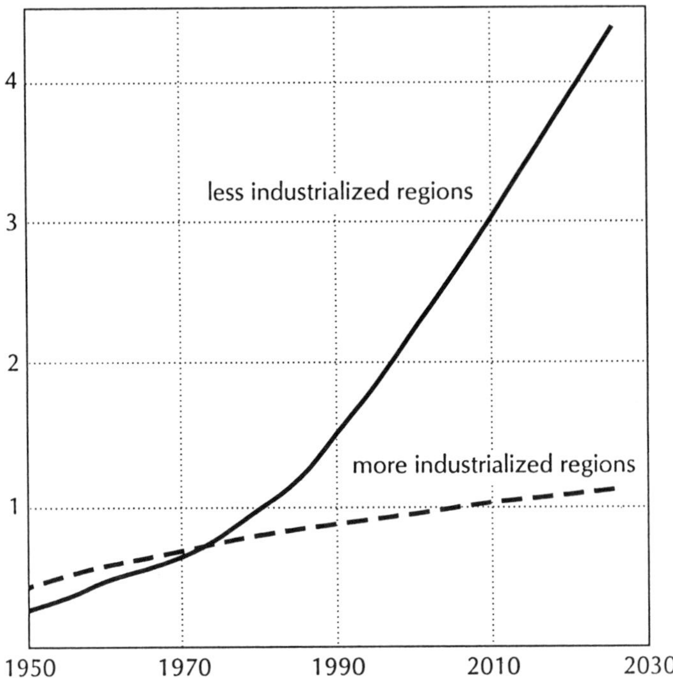

Total urban population is expected to increase exponentially in the less industrialized regions of the world, but almost linearly in the more industrialized regions. Average doubling time for city populations in less industrialized regions has been 20 years—faster than population growth as a whole. (*Sources: United Nations; Population Reference Bureau.*)

That is exponential growth, doubling and redoubling and doubling again. Nearly everyone is surprised by it, because most people think *linearly* and think of growth as a linear process. A quantity grows linearly when it increases by a constant amount in a constant time period. If a construction crew produces a mile of highway each week, the length of the road grows linearly. If a child puts $10 a year in a piggy bank, his or her savings increase linearly. In linear growth *the amount of increase is constant in a given time period.* It is not affected by the length of the road already built or the amount of money already in the bank.

Figure 2-3 LINEAR VERSUS EXPONENTIAL GROWTH OF SAVINGS

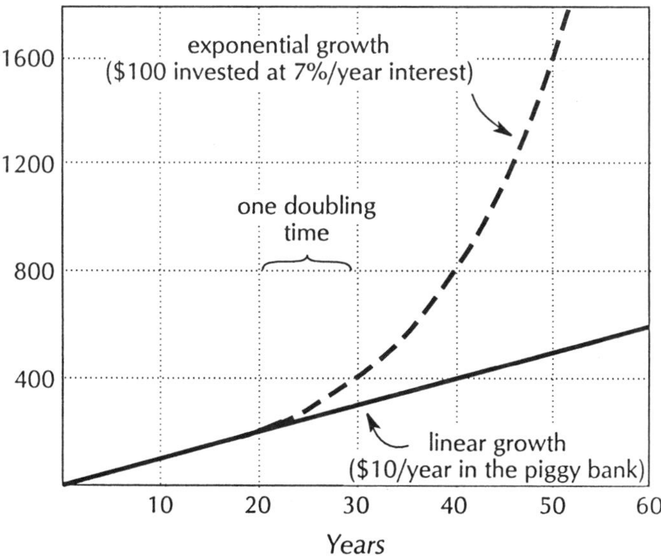

If a child puts $10 each year in a piggy bank, the savings will grow linearly, as shown by the lower curve. If, starting in year 10, the child invests $100 at 7% interest, that $100 will grow exponentially, with a doubling time of 10 years.

A quantity grows *exponentially* when *its increase is proportional to what is already there.* A colony of yeast cells in which each cell divides into two every 10 minutes is growing exponentially. For each single cell after 10 minutes there will be two cells. After the next 10 minutes there will be four cells, 10 minutes later there will be eight, then sixteen, and so on. The more yeast cells there are, the more new ones can be made.

If the child invested $100 at 7% per year interest (and let the interest income accumulate in the account), the invested money would grow exponentially. It would mount up much faster over the long run than would the linearly increasing stock in the piggy bank (see Figure 2-3). The first year's interest will be 7% of $100 or $7, making a total of $107 in the account. The next year's interest will be 7% of $107, which is $7.49, bringing the total to $114.49. One year later the interest on that amount will be $8.01, and the total will be $122.50. By the tenth year the account will have grown to $201.37. And so forth.

The *percent* added each year to a bank account or each 10 minutes to a yeast colony is constant, but the *amount* added is not. It gets larger and larger as the total accumulation of money or yeast increases.

The surprising consequences of exponential growth have fascinated people for centuries. There is an old Persian legend about a clever courtier who presented a beautiful chessboard to his king and requested that the king give him in exchange 1 grain of rice for the first square on the board, 2 grains for the second square, 4 grains for the third, and so forth.

The king readily agreed and ordered rice to be brought from his stores. The fourth square on the chessboard required 8 grains, the tenth square took 512 grains, the fifteenth required 16,384, and the twenty-first square gave the courtier more than a million grains of rice. By the fortieth square a million million rice grains had to be piled up. The payment could never have continued to the sixty-fourth square; it would have taken more rice than there was in the whole world.

A French riddle for children illustrates another aspect of exponential growth—the apparent suddenness with which an exponentially growing quantity approaches a fixed limit. Suppose you own a pond on which a water lily is growing. The lily plant doubles in size each day. If the plant were allowed to grow unchecked, it would completely cover the pond in 30 days, choking off the other forms of life in the water. For a long time the lily plant seems small, so you decide not to worry about it until it covers half the pond. On what day will that be?

On the twenty-ninth day. You have just one day to act to save your pond.[3] (On the twenty-fifth of the month the plant covers just 1/32nd of the pond; on the twenty-first it covers just 1/512th of the pond. For most of the month the plant, though it is steadily doubling, is invisible or inconsequential. You can see how exponential growth, combined with inattention, can lead to overshoot!)

A quantity that is growing according to a pure exponential growth equation doubles again and again, and each doubling takes the same time as the doubling before. In the case of the lily plant the doubling time is one day. Money left in a bank at 7% interest will double every

10 years. There is a simple relationship between the interest rate, or rate of growth in percentage terms, and the time it will take a quantity to double. The doubling time is approximately equal to 70 divided by the growth rate, as illustrated in Table 2-1.

Table 2-1 DOUBLING TIMES

Growth rate (% per year)	*Doubling time (years)*
0.1	700
0.5	140
1.0	70
2.0	35
3.0	23
4.0	18
5.0	14
7.0	10
10.0	7

Here is a hypothetical example of how doubling times work. Nigeria had a population in 1990 of 118 million, and its population growth rate was 2.9% per year. The doubling time for this rate of growth is 70 divided by 2.9 or 24 years. If its current population growth rate continued unchanged into the future, Nigeria's population would follow a pattern like the one illustrated in Table 2-2.

Table 2-2 NIGERIA'S POPULATION WITH CONTINUED EXPONENTIAL GROWTH

Year	*Population (millions)*
1990	118
2014	236
2038	472
2062	944
2086	1888

A Nigerian child born in 1990 and living for 70 years would see the population multiply almost eightfold. Near the end of the next century there would be over 1.8 billion Nigerians, 16 for every one in 1990. By the year 2086 almost three times as many people would live in Nigeria as lived in all of Africa in 1990!

The only reason for doing a calculation like this is to become convinced that such a future can never happen. Exponential growth simply cannot and will not go on very long.

So why is it going on now? And what is likely to stop it?

Things That Grow Exponentially

Exponential growth happens for one of two reasons: because a growing entity reproduces itself out of itself, or because a growing entity is *driven by* something that reproduces itself out of itself.

All living creatures from bacteria to people fall under the first category. New creatures are produced by other creatures. The more creatures there are, the more new ones can be made.

We illustrate the system structure of a self-reproducing population with a diagram like this:

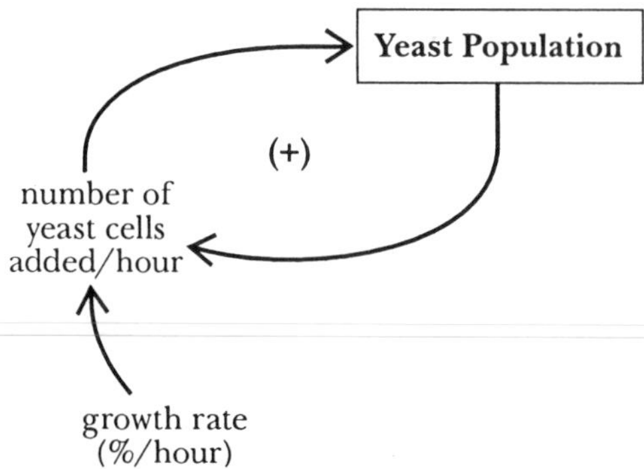

The box around the yeast population indicates that it is a *stock*—an accumulation of yeast, the result of past multiplications. The arrows indicate causation or influence, which may be exerted in many ways. In this diagram the top arrow means that new yeast flow into and increase the stock. The bottom arrow means that the size of the stock governs the generation of new yeast. The greater the population, the more new yeast cells will be made, as long as nothing happens to stop the growth.

The (+) sign in the middle of the loop means that the two arrows together make up a *positive feedback loop*. A positive feedback loop is a chain of cause-and-effect relationships that closes in on itself so that a change in any one element in the loop will change the original element even more in the same direction. An increase will cause a further increase; a decrease will cause a further decrease.

In this sense "positive" doesn't necessarily mean good. It simply refers to the *reinforcing* direction of the causal influence around the loop. (Similarly, negative feedback loops, which we'll discuss in a moment, aren't necessarily bad. In fact they're often stabilizing. They are negative in the sense that they *counteract* or *reverse* causal influence around the loop.)

A positive feedback loop can be a "virtuous circle," or a "vicious circle," depending on whether the type of growth it produces is wanted or not. Positive feedback causes the exponential growth of yeast in rising bread, of a pest outbreak in an agricultural crop, of a cold virus in your throat, and of money in an interest-bearing bank account. Whenever a positive feedback loop is present in a system, that system has the potential to produce exponential growth or exponential decline.

The presence of a positive growth loop doesn't mean that a population of yeast, people, pests, or money necessarily will grow exponentially; it only means that it has the *structural capacity* to do so. The actual growth rate will be influenced by many things, such as nutrients (in the case of yeast), interest rate (in the case of money), temperature and the presence of other populations (in the case of pests), and, in the case of human beings, incentives, disincentives, goals, and purposes. The actual rate of growth may vary greatly over time or place. A population's structural capacity for growth can be held in check either by an outside factor or by self-restraint. But population growth, when it does occur, is exponential, until something stops it.

Something else that can grow exponentially is *industrial capital,* by which we mean the machines and factories that generate other machines and factories. A steel mill can make the steel to build another steel mill, a nuts-and-bolts factory can make the nuts and bolts that hold together machines that make nuts and bolts. More factories make even more factories possible, in the interconnected, self-supplying, cross-supplying way that the industrial economy has evolved.

It's not an accident that the world has come to expect an economy to grow by a certain percentage of itself—3% or 5% or so—each year. That's an expectation of exponential growth. It can be realized simply because capital can create itself out of itself. An economy will grow exponentially whenever the self-reproduction of capital is unconstrained by consumer demand, by labor availability, by raw materials or energy, by investor confidence, by incompetence, by any of the hundreds of factors that can limit the operation of a complex production system. Like population, capital has the system *structure* (a positive feedback loop) to produce the *behavior* called exponential growth. But capital has other feedback loops influencing it too and other possible behaviors. Everyone knows that economies don't always grow. But they have a strong tendency to grow, and most of them do grow, whenever possible.

Population and capital are engines of growth in the industrialized world. Other quantities, such as food production, resource use, and pollution, tend to increase exponentially not because they multiply themselves, but because they are *driven by* population and capital. There is no self-generation, no positive feedback loop to cause pesticides in groundwater to create more pesticides, or coal to breed underground and produce more coal. Growing 2 million tons of wheat does not in itself make it easier to grow 4 million tons, unless there has been learning or a technical development in the process. At some point as limits are reached, each doubling of food output or mined materials is not easier but more difficult than the doubling before.

Therefore food production and resource and energy use have been growing not through their own structural capacity, but because an exponentially growing population has been demanding more food and materials and energy and so far has been successful at producing them. Similarly pollution and waste have been growing not because of their own internal positive feedback processes, but because they are driven by the rising quantities of materials moved and energy consumed by the growing human economy.

Population and capital are capable of exponential growth, and as they grow they demand and facilitate the growth of material and energy throughputs and pollution and waste emissions. That is not an arbitrary assumption, it is a fact. It is a structural fact—the mechanisms by which it happens are understood. And it is an observed fact—the human population and capital plant and the energy and material flows that sustain them have been growing vigorously, with only a few brief interruptions, for centuries.

World Population Growth

In the year 1650 the human population numbered around 0.5 billion. It was growing at about 0.3% per year, corresponding to a doubling time of nearly 250 years.

By 1900 the population had reached 1.6 billion and was growing at 0.5% per year, a doubling time of 140 years.

By the year 1970 the population totalled 3.6 billion and the rate of growth had increased to 2.1% per year. That was not only exponential growth, it was superexponential—the rate of growth was itself growing. It was growing for a happy reason: death rates were falling. Birth rates were also falling, but much more slowly. Therefore the population surged.

Between 1971 and 1991 death rates continued to fall, but birth rates on average fell slightly faster (Figure 2-4). While the population rose from 3.6 billion to 5.4 billion, the *rate* of growth fell from 2.1% to 1.7%.[4]

That's a significant change, but it does not mean that population growth is anywhere close to leveling off. In fact more people were added to the world in 1991 than in any year ever before. Table 2-3 shows why.

Table 2-3 ADDITIONS TO WORLD POPULATION, 1971 AND 1991

Year	*Population (millions)*	×	*Growth rate (per year)*	=	*People added (millions)*
1971	3600	×	2.1%	=	76
1991	5400	×	1.7%	=	92

The population *growth rate* has not dropped as fast as the population *base* has grown. Therefore the number of people added each year continues to increase. Growth is still exponential, though at a slightly lower rate. The 92 million added in 1991 is equivalent to adding in that year the total populations of Germany plus Switzerland plus Austria—or about six New York Cities—or more accurately, since 90% of the increase takes place in the Third World, it is equivalent to adding in one year the total populations of Mexico plus Honduras—or about eight Calcuttas. Even under extremely optimistic projections about further drops in birth rates, an enormous increase of population is still ahead, especially for the less-industrialized countries (Figure 2-5).

The central feedback structure that governs the population system is shown below.

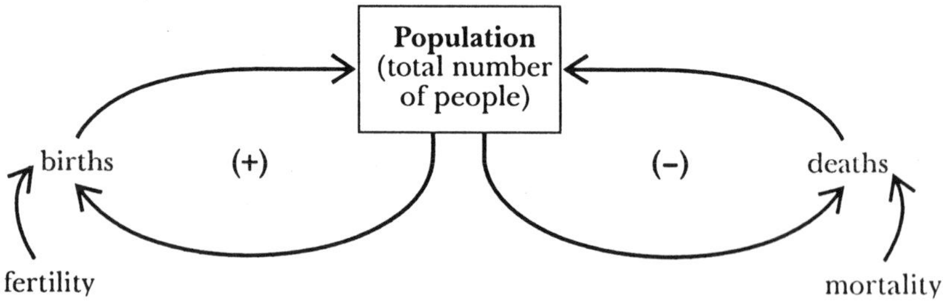

Figure 2-4 WORLD DEMOGRAPHIC TRANSITION

Births & deaths per 1000 per year *Population (billions)*

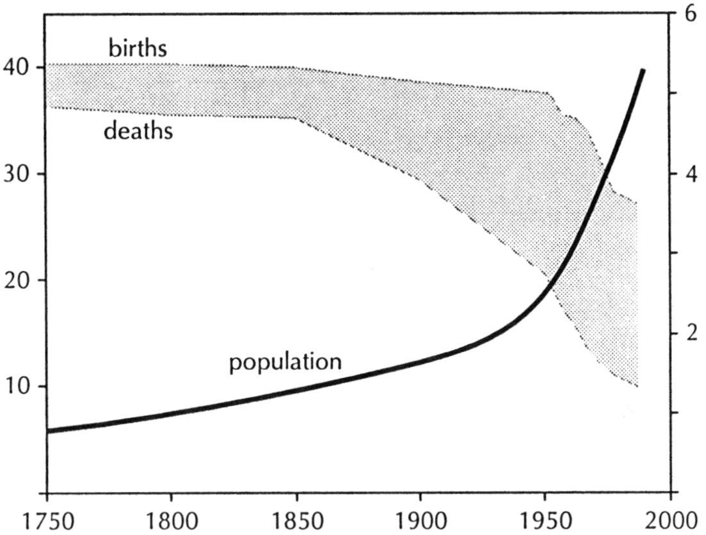

The shaded gap between births and deaths shows the rate at which the population grows. Until about 1970 the average human death rate was dropping faster than the birth rate, and the population growth rate was increasing. Since 1970 the average birth rate has dropped slightly faster than the death rate. Therefore the rate of population growth has decreased somewhat—though the growth continues to be exponential. (*Source: United Nations.*)

On the left is the positive loop that accounts for the exponential growth. The larger the population, the more babies will be born each year. The more babies, the larger the population. After a delay while those babies grow up and become parents, even more babies can be born, swelling the population still further.

On the right is another feedback loop that governs population growth. It is a *negative feedback loop.* Whereas positive loops generate runaway growth, negative feedback loops tend to regulate growth, to hold a system within some acceptable range, or to return it to a stable state. A negative feedback loop propagates the consequences of a

Figure 2-5 WORLD ANNUAL POPULATION INCREASE

Millions of people added each year

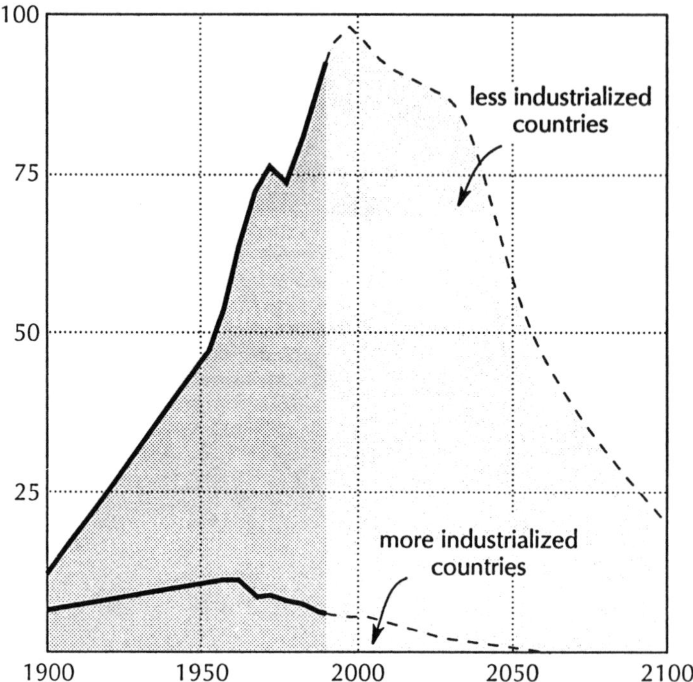

The number of people added to the world population each year has increased enormously and is projected to go on increasing for another decade under the World Bank's forecasts. Those forecasts are very optimistic; they assume rapid drops in birth rates in the less industrialized countries. (*Sources: United Nations; E. Bos et al.*)

change in one element around the circle until they come back to change that element in a direction *opposite* to the initial change.

The number of deaths each year equals the total population times the average mortality—the average probability of death at each age. The number of births equals the total population times the average fertility. The growth rate of a population is equal to its fertility minus its mortality. Of course human fertility and mortality are not at all constant. They depend upon economic, environmental, and demographic factors such as income, education, health care, family planning technologies, religion, pollution, and the population's age structure.

Therefore the two simple feedback loops pictured above can produce a number of different dynamic behaviors. If fertility is higher than mortality, the population will grow exponentially.

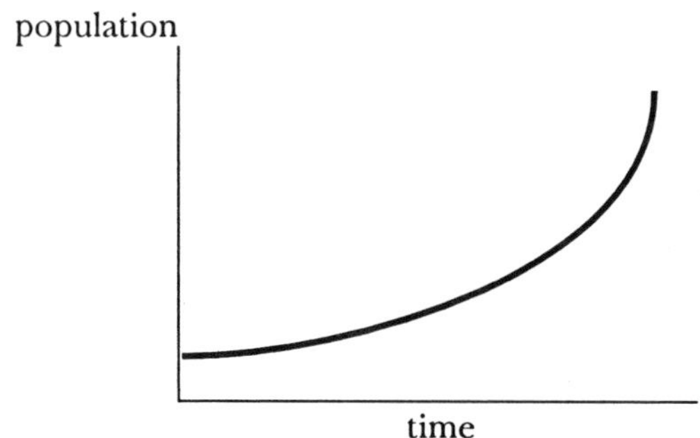

If mortality is higher than fertility, the population will decline toward zero.

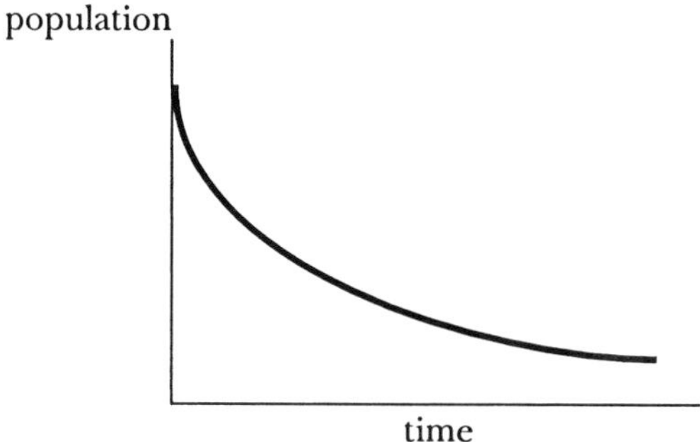

If fertility just equals mortality, births will equal deaths and the population size will stay constant, though there will be a continuous turnover, a flow of new people replacing old. This condition of steady flow is called *dynamic equilibrium.*

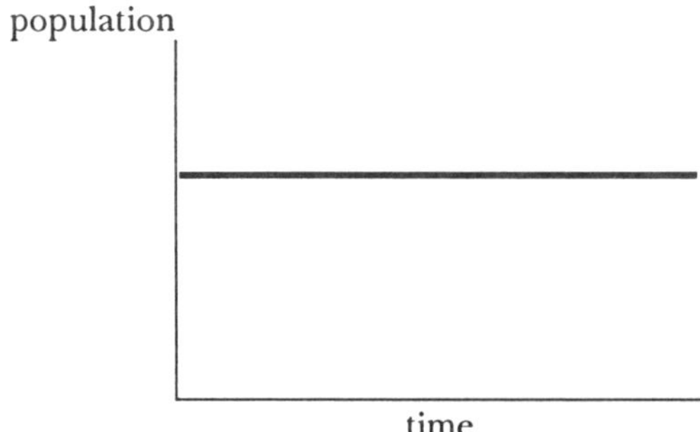

Actual populations can exhibit any of those behaviors over time and do. The fertility and mortality combinations of human populations are as varied as the cultures and histories of the many nations and ethnic groups of the world. But within the variety there are a few regularities:

- Some of the least industrialized populations, such as many in Africa, still have relatively high mortality and even higher fertility. Their rate of population growth is 2% to 3% per year, and it may increase as mortality declines.
- Populations at an intermediate level of industrialization, such as those of Brazil, Indonesia, Thailand, and Egypt, typically have low mortality while their fertility is still high but also decreasing. They are growing at moderate to rapid rates (1% to 4% per year).
- Most highly industrialized populations, like those in North America, Japan, and Europe have low mortality, low fertility, and slow (less than 1% per year) growth rates. The birth rates of a few European populations have recently dropped below their death rates, so those populations are very slowly declining.

Demographers have postulated from this set of patterns a theory called the *demographic transition*. According to this theory, at low levels of industrialization both fertility and mortality are high, and population growth is slow. As nutrition and health services improve, death rates fall. Birth rates lag behind by a generation or two, opening a gap between fertility and mortality that produces rapid population growth. Finally, as people's lives and lifestyles evolve into a fully industrial mode, birth rates fall too, and the population growth rate slows again.

The actual demographic experiences of six countries are shown in Figure 2-6. You can see from this figure that birth and death rates in the long-industrialized countries such as Sweden fell very slowly. The gap between them was never very high; the population never grew at more than 2% per year. Over the entire demographic transition the populations of most countries of the North grew by at most a factor of 5.

In the countries of the South death rates fell much later and much faster. A large gap opened up between birth and death rates, and these countries are experiencing rates of population growth much faster than any the North ever had to deal with (except for North America, which absorbed for awhile high rates of immigration from Europe). The populations of many presently industrializing countries of the South have already grown by factors of 5 to 10 and are still growing rapidly. Their demographic transitions are far from over, and rapid population growth may itself be slowing down those transitions.

Demographers argue about what actually *causes* the demographic transition, especially the crucial fall in the birth rate. The driving factor is something more complicated than simple income. Figure 2-7 shows, for example, the correlation between GNP per capita and birth rates in various countries of the world. Clearly there is *some* relationship between economic output per capita and birth rate. Just as clearly there are major exceptions. China and Sri Lanka, for example, have anomalously low birth rates for their level of income. Several Middle Eastern countries have anomalously high birth rates for theirs.

The factors believed to be most *directly* important in lowering birth rates are not so much the average national level of income, but the extent to which that income actually changes the lives of families, and especially the lives of women. More important than GNP per capita are factors such as education and employment (especially for women), family planning, and reduction of infant mortality. China, Sri Lanka, Costa Rica, Singapore, Thailand, Malaysia, and several other countries have demonstrated that all these birth-rate–reducing factors can be provided

Figure 2-6a Demographic Transitions in Industrialized Countries

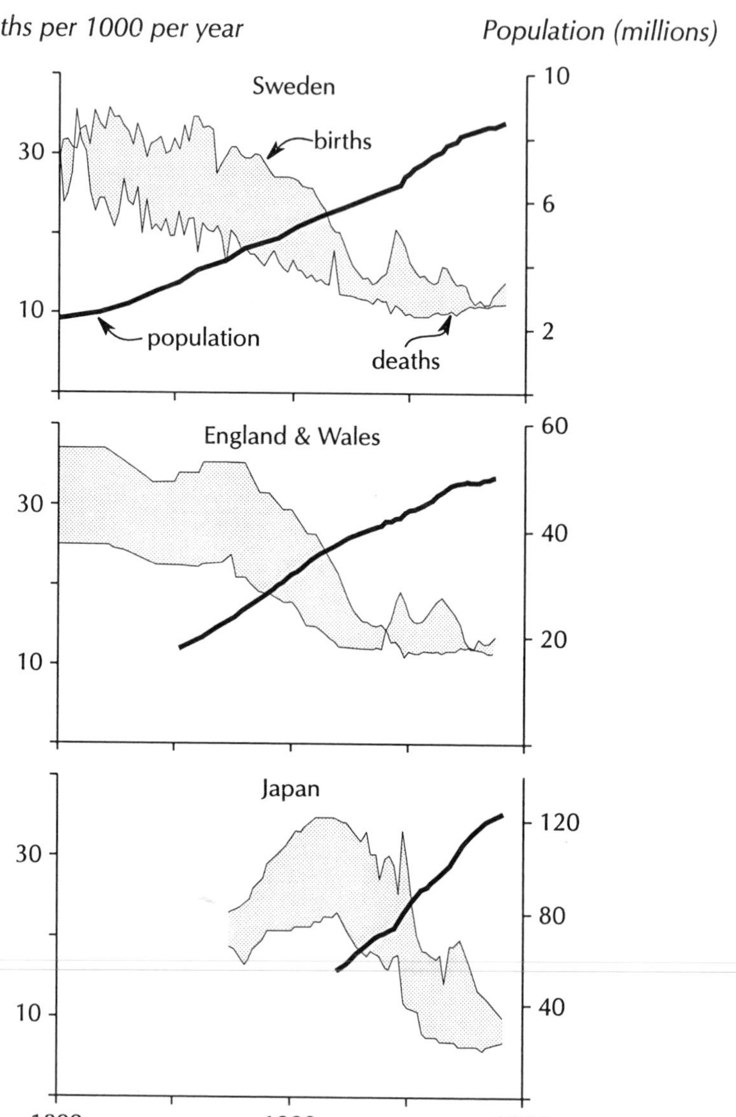

In the demographic transition a nation's death rate falls first, followed later by its birth rate. Sweden's demographic transition occurred over almost 200 years, with the birth rate remaining rather close to the death rate. During this time Sweden's population increased less than fivefold. Japan is an example of a nation that will effect the transition in less than a century. The less-industrialized countries have

Figure 2-6b DEMOGRAPHIC TRANSITIONS IN LESS-INDUSTRIALIZED COUNTRIES

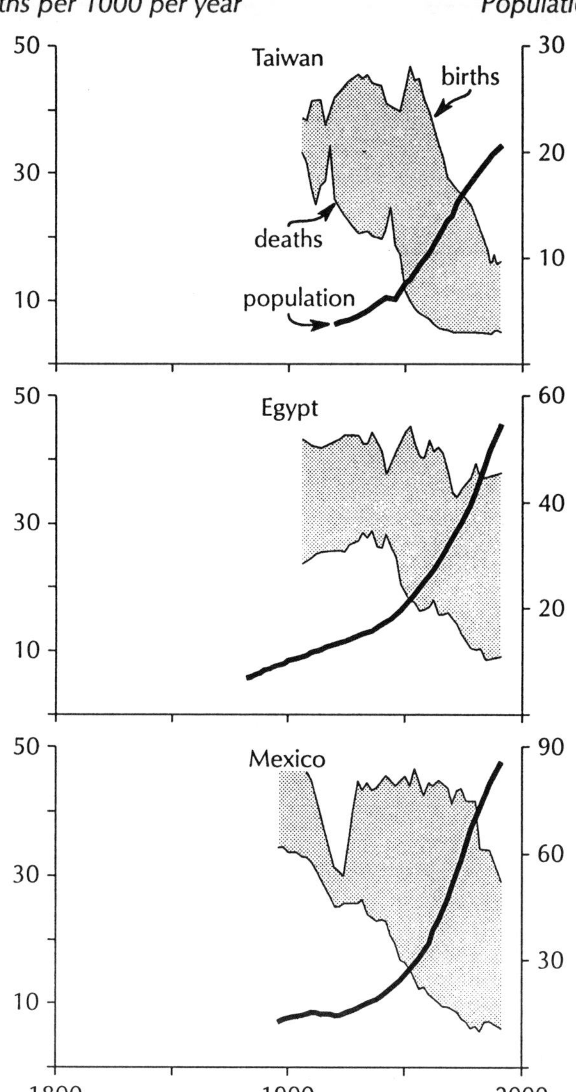

less time to accomplish this shift, and the gaps between their birth and death rates are larger than any that prevailed in the long-industrialized countries. (*Sources: United Nations; R. A. Easterlin; J. Chesnais; N. Keyfitz; Population Reference Bureau; U.K. Office of Population Census and Surveys.*)

Figure 2-7 BIRTH RATES AND GNP PER CAPITA IN 1989

Births per 1000 per year

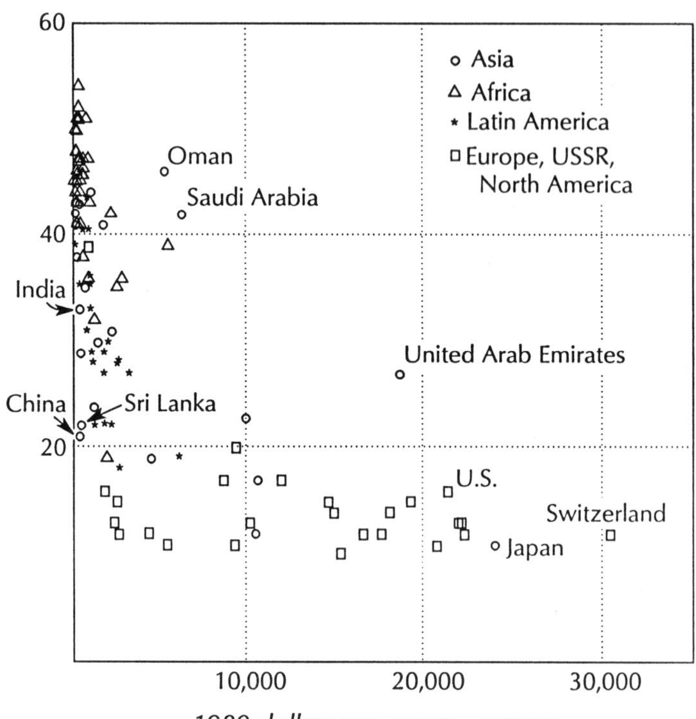

As a society becomes more wealthy, the birth rate of its people tends to decline. All of the poorest nations experience birth rates between 20 and 50 per thousand people per year. None of the richest nations have birth rates above 20 per thousand per year except the oil-rich states of the Middle East. (*Sources: Population Reference Bureau; CIA.*)

to most families at low cost, but only if a society decides to allocate its resources that way.

Industrial growth does not guarantee improvements in actual human welfare or reductions in the growth rate of a population. But it can certainly help. Therefore, it is doubly important to understand the causes and consequences of industrial growth.

World Industrial Growth

Public discussions of economic matters are full of confusions, many of which come from failing to distinguish between money, the real things money stands for, and the different functions those real things play in an economy. We need to make those distinctions carefully here. Figure 2-8 shows how we represent the physical economy in World3, and how we will talk about it in this book.

Industrial capital means here the actual hardware, the physical machines and factories that produce manufactured products. (With the help, of course, of labor, energy, raw materials, land, water, technology, finance, management, and the services of the natural ecosystems of the planet. We will come back to these cofactors of production, especially energy, raw materials, land, water, and the planet's services, in the next chapter.) We call the continuous stream of products that is made by industrial capital *industrial output.*

Some industrial output is intended for *final consumption*—cars, clothing, radios, refrigerators, houses.

Some industrial output takes the form of drills, oil wells, mining equipment, pipelines, tankers. All that is *resource-obtaining capital,* which produces the output stream of resources necessary to allow all the other forms of capital to function.

Some industrial output is *agricultural capital*—tractors, barns, irrigation systems, harvesters—which produce *agricultural output,* mainly food.

Some industrial output is equipment or buildings for hospitals, schools, banks, retail stores. That is *service capital.* Service capital produces its own stream of output—health care, education, and so on.

And finally some industrial output is more industrial capital, which we call *industrial investment*—more steel mills, electric generators, lathes and other machines, which increase the stock of industrial capital to allow more output in the future.

So far everything we have mentioned here is real, physical stuff, not money. The role of money is to convey information about relative costs and values of that stuff (values as assigned by the producers and consumers who have power in the market). Money flows mediate and moti-

Figure 2-8 Flows of Capital in the Economy of World3

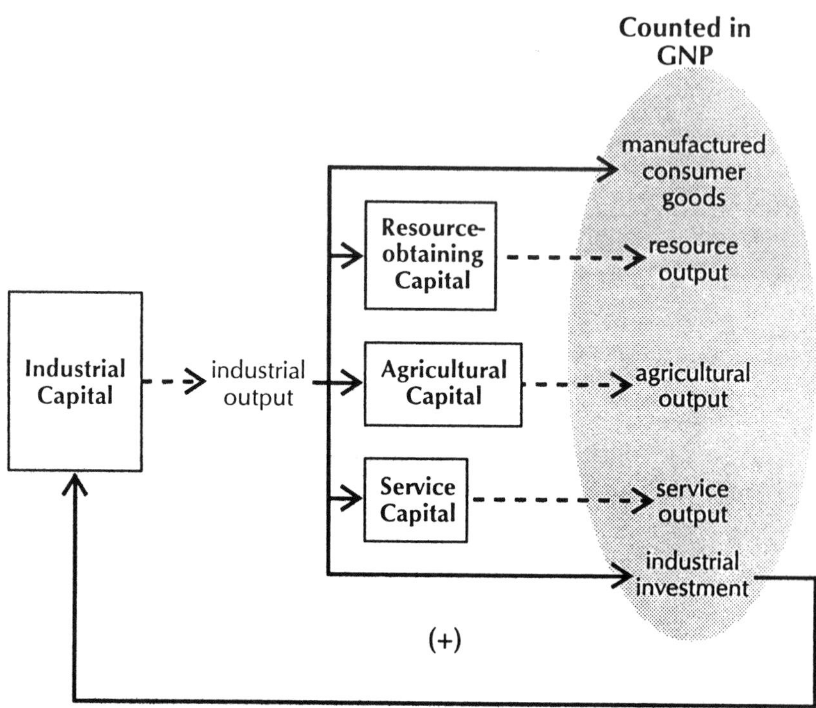

The production and allocation of industrial output is a central influence on the behavior of the economy in World3. The amount of industrial capital determines how much industrial output can be produced each year. This output is allocated among five sectors in a way that depends on the goals and needs of the population. Some industrial output is consumed, some is allocated to the resource sector to secure raw materials. Some industrial output goes to agriculture to develop land and raise land yield. Some industrial output is invested in social services and the rest is invested in industry to raise the industrial capital stock further.

vate the flows of physical capital and products. The annual money value of all outputs of final goods and services shown in Figure 2-8 is the Gross National Product (GNP). Our focus here will be less on money flows than on physical flows, because physical flows, not money flows, are limited by the support systems of the earth. We will refer to GNP in various figures and tables, because the world's economic data are kept in money terms, not physical terms. But our interest is in what GNP stands for: material flows of capital, industrial goods, services, resources, and agricultural products.

We have already said that industrial capital is something that can grow exponentially by its own self-generation. The feedback structure representing that self-generation is similar to the one we drew for the population system.

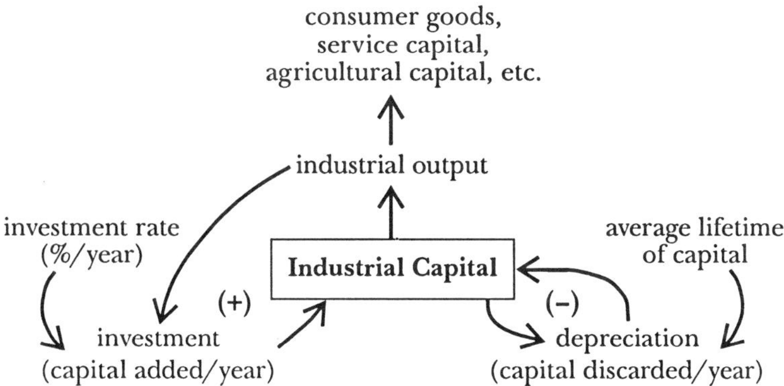

A given amount of industrial capital (factories, trucks, computers, power plants, etc.) can produce a certain amount of manufactured output each year, as long as labor, energy, raw materials, and other necessary inputs are sufficient. Some percent of each year's production is investment—looms, motors, conveyer belts, steel mills—which goes to increase the capital stock and thereby expand the capacity for production in the future. The percent invested is variable, as human fertility is variable, depending on human decisions and economic constraints.

The positive feedback loop of investment is shown on the left in the diagram above. More capital creates more output, some of it investment, and more investment creates more capital. The new, larger capital stock can generate even more output, and so on. There are delays in this feedback loop, since the planning, financing, and construction time for a major piece of capital equipment, such as a railroad, electric generating plant, or refinery, can take years or even decades.

Capital, like population, has a "death loop" as well as a "birth loop." As machines and factories wear out or become technically obso-

Figure 2-9 U.S. GNP BY SECTOR

Billions of 1982 dollars per year

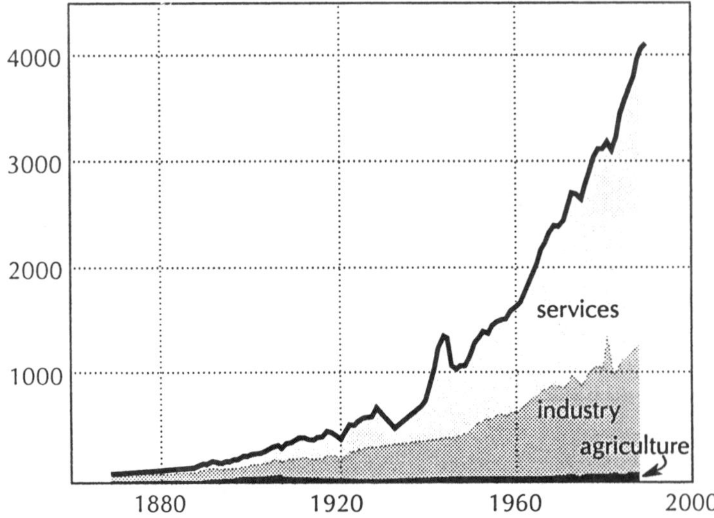

The history of the distribution of the U.S. GNP among services, industry, and agriculture shows the transition to a service economy. Note that although services assume the largest share of the economy, the industrial and agricultural sectors still continue to expand in absolute terms. (*Sources: U.S. Bureau of the Census; U.S. Council of Economic Advisors.*)

lete, they are shut down, dismantled, and discarded. The rate of capital depreciation is analogous to the death rate in the population system. The more capital is present, the more there is to wear out each year, so the less there will be the next year, unless the inflow of new investment is sufficient to replace depreciated capital.

Since it is driven by similar feedback loops, capital is capable of the same three basic behavior modes as population: exponential growth, exponential decline, and dynamic equilibrium. Just as populations undergo a demographic transition during the process of industrialization, economies also undergo a long-term transition. Preindustrial economies are primarily agriculture and service economies. As the capital growth loop starts operating, all economic sectors grow, but the industrial sector grows fastest for a while. Later, when the industrial base has been built, further growth takes place primarily in the service sector (see Figure 2-9).

Highly industrial economies are sometimes said to be evolving into "service economies," but in fact they continue to require a substantial agricultural and industrial base. Hospitals, schools, banks, stores, restaurants, and tourist facilities are all part of the service sector. If you have ever watched the delivery trucks bringing them food, paper, fuels, and equipment, or the trucks hauling their solid waste away, you know that the modern service sector rests solidly on a base of agricultural and industrial production.

As in the population system, the positive loop in the capital system is strongly dominant in the world today. Industrial capital has been growing exponentially, faster than the population. Over the twenty years from 1970 to 1990, industrial output grew by nearly 100% (as shown in Figure 1-2). That growth would have produced on average twice as much industrial output per person in the world as there was twenty years ago if population had been constant, but because of population growth the average industrial output per person only grew by about one-third.

More Poverty, More People, More Poverty

If capital grows faster than population, that should mean, according to the theory of the demographic transition, that the rising material standard of living of the world's people is bringing down the population growth rate.

To some extent and in some places that is exactly what is happening. But neither economic growth nor its demographic response is taking place as quickly as it might, and in some parts of the world both are actually going backward: economic welfare is falling and population growth rates are stagnant or rising. That is because of the way growth in the industrial economy is distributed.

Most economic growth takes place in the already industrialized countries. Figure 2-10 shows GNP per capita growth curves for some of the world's most populous nations. They illustrate how economic growth systematically continues to occur more in the rich countries

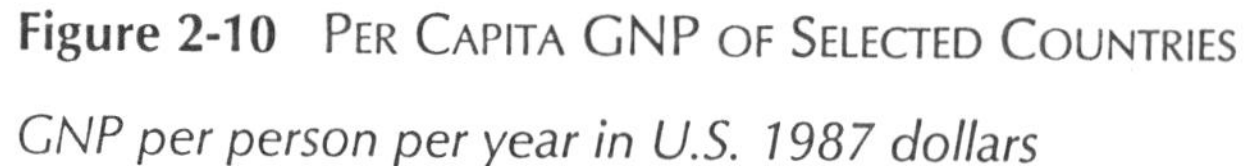

Figure 2-10 PER CAPITA GNP OF SELECTED COUNTRIES

GNP per person per year in U.S. 1987 dollars

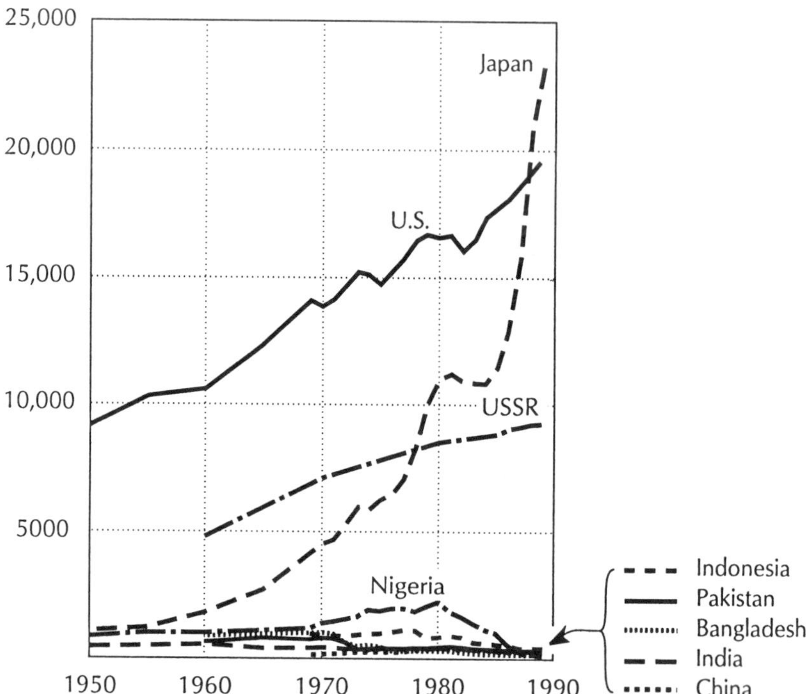

Economic growth takes place primarily in the nations that are already rich. The five countries of Indonesia, China, Pakistan, India, and Bangladesh together contain almost half the world's population. Their per capita GNP barely rises off the axis when it is plotted together with the GNP per capita of the wealthier nations. (*Sources: World Bank; CIA.*)

than in the poor ones. The 1991 *World Bank Development Report* lists forty countries in the less-industrialized world whose per capita incomes actually fell over the decade of the 1980s. These countries contain more than 800 million people, almost three times the population of North America and nearly one-sixth the population of the world.[5]

There are many reasons for the economic stagnancy of poor nations and populations, some of which have to do with systematic injustice, oppression, and neglect from those who are not poor, and some of which have to do with the lack of training, information, opportunity, health, and management skills in the poor countries. There are also reasons that have to do with the simple structure of the population and capital system as we have described it in this chapter.

It is much easier for rich populations to save, invest, and multiply their capital than for poor ones to do so, not only because of the greater power of the rich to control market conditions, purchase new technologies, and command resources, but also because centuries of past growth have built up in rich countries a large stock of capital that can multiply itself yet more. Most basic needs are met, so relatively high rates of saving and investment for the future are possible without impoverishing the present. The lower population growth in the richer countries permits output to be allocated more toward industrial investment and less toward the service investment needed to meet the health and education needs of a rapidly expanding population.

In poor countries capital growth has a hard time keeping up with population growth for many reasons—because investable surplus is siphoned off to foreign investors, to the luxury of local elites, to debt repayments, or to exorbitant militarization—and because there is too much poverty, technical inefficiency, or mismanagement to generate an investable surplus in the first place. The population is stuck in a pattern of growing bigger without growing richer.

The system structure that links together population and capital is such that the most common behavior of the world system is the one captured in the old saying "the rich get richer and the poor get children." It is no accident that the system produces that behavior; it is structured to do so, and will continue to, unless that structure is deliberately changed. Population growth slows industrial capital growth by creating rising demand for schools, hospitals, resources, and basic consumption, thereby drawing industrial output away from industrial investment. Poverty perpetuates population growth by keeping people in conditions where they have no education, no health care, no family planning, no choices, no way to get ahead except to have a large family and hope the children can bring in income or help with family labor.

International gatherings can break into passionate arguments about which arrow in this feedback loop is most important: poverty causes population growth, or population growth causes poverty.

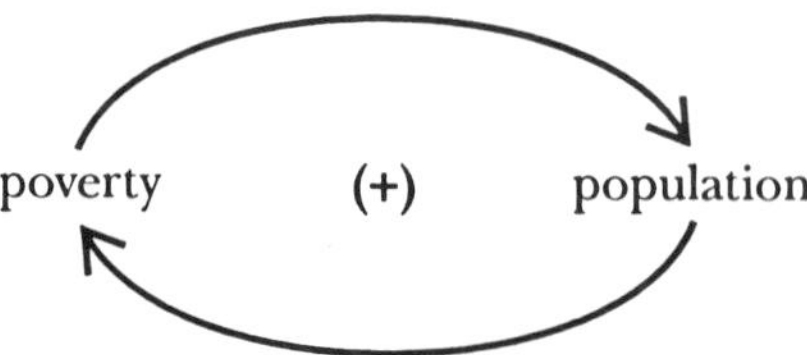

In fact both arrows are operative, and they can reinforce each other in a positive feedback process that grinds downward, forming a trap that keeps the poor poor and the population growing. One consequence of this trap is shown in Figure 2-11. Food production in every part of the Third World has increased greatly over the past twenty years. In most places it has doubled or tripled. But because of rapid population growth, food production per person has barely improved, and in Africa it has steadily decreased. During the period 1985 to 1989 food production per capita declined in 94 nations.[6]

The graphs in Figure 2-11 show a double tragedy. The first tragedy is a human one. An agricultural achievement, a tremendous increase in food production has been absorbed not in feeding hungry people more but in feeding more hungry people. The second tragedy is environmental. The increase in food production was won at great cost to the earth, and that cost will make future increases more difficult. Because of the poverty-population trap, an agricultural success has resulted primarily in more people and more deserts.

Any positive feedback loop that grinds a system down, however, can be turned around to work in the other direction. More prosperity, widely distributed, can lead to slower population growth, which can lead to more prosperity. With enough investment sustained for a long enough time, with fair pricing for products and fair market conditions, with the increased output allocated to the poor and especially to the education and employment of women, a population can lift itself out of poverty.

That process has happened spectacularly in some places. But the majority of the world's people are still fighting to get out of the poverty-population trap. And, as Figures 2-10 and 2-11 indicate, the pattern of economic growth as it has occurred over the past few decades is not helping them.

The human mind has a tendency to classify things as "bad" or "good" and to keep those classifications fixed permanently. For generations both population growth and capital growth have been classified as "good." On a lightly populated planet with abundant resources there were excellent reasons for that evaluation. Now, with dawning awareness of ecological limits, some people want to classify all material growth as "bad."

But the task of managing in the presence of ecological limits demands of the human mind greater subtlety, more careful classification. Poorer people desperately need more food, shelter, and material goods. Wealthier people, in a different kind of desperation, try to use material growth to satisfy other needs, which are also very real but are in fact nonmaterial–needs for acceptance, self-importance, community, identity. It makes no sense at this time of rapid growth on a finite planet to talk about growth with either unquestioning approval or unquestioning disapproval. Instead it is necessary to ask: Growth of what? For whom? For how long? At what cost? Paid by whom? What is the real need here, and what is the most direct and efficient way for those who have the need to satisfy it?

Those questions can point the way toward a society that is sufficient and equitable. Other questions will point the way toward a society that is sustainable. How many people can be provided for on this planet? At what level of material consumption? For how long? How stressed is the physical system that supports the human population, the human economy, and all other species? How resilient is that support system to what kinds and quantities of stress?

To answer those questions, we must look not at growth, but at limits to growth.

Figure 2-11 REGIONAL FOOD PRODUCTION

Index (1952–56 = 100)

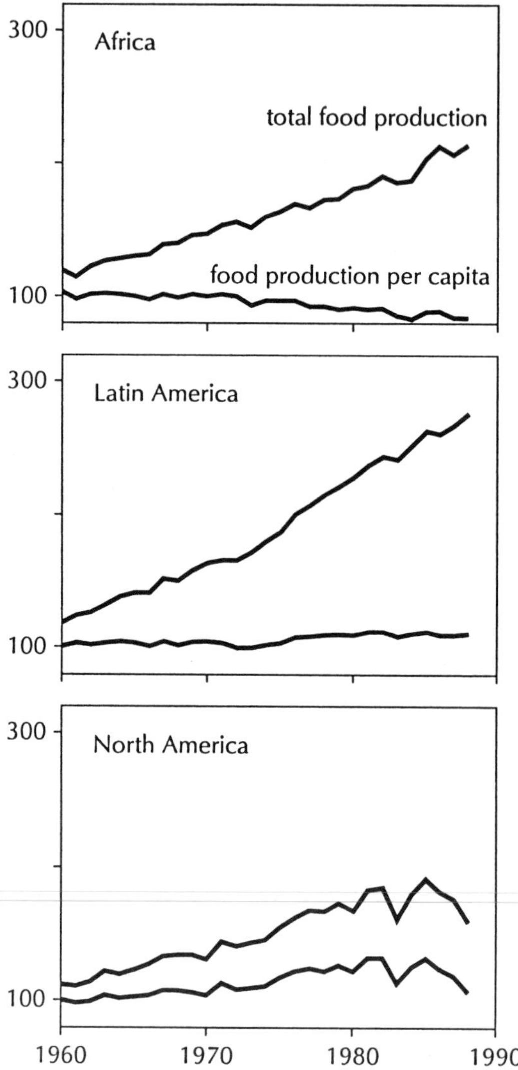

Total food production has doubled or tripled in the past 30 years in the regions of the world where hunger is greatest, but food per person has scarcely changed in those areas, because population has grown almost as fast. (*Source*: *Food and Agriculture Organization*.)

Figure 2-11 (continued)

Index (1952–56 = 100)

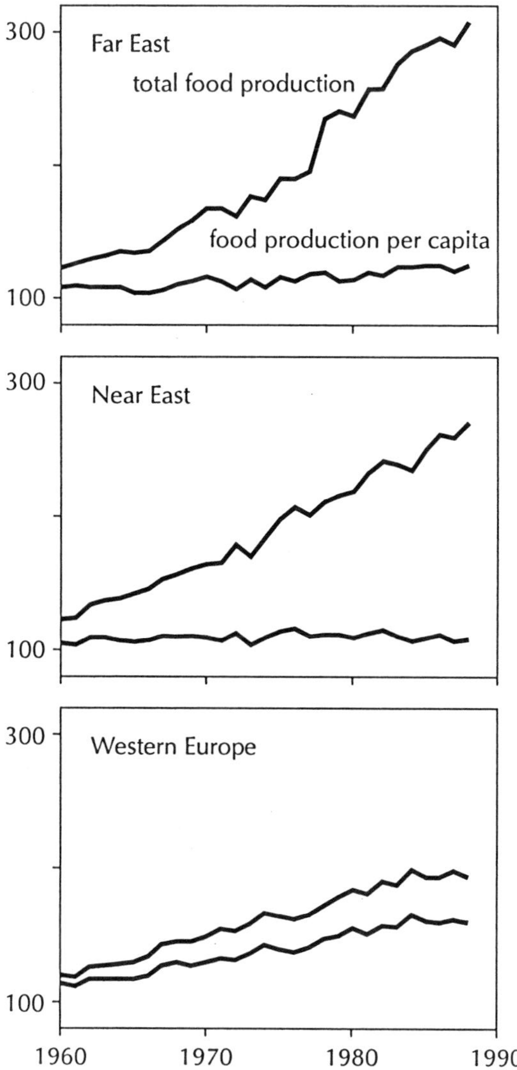

chapter 3

The Limits: Sources and Sinks

Many present efforts to guard and maintain human progress, to meet human needs, and to realize human ambitions are simply unsustainable–in both the rich and poor nations. They draw too heavily, too quickly, on already overdrawn environmental resource accounts. . . . They may show profits on the balance sheets of our generation, but our children will inherit the losses.

World Commission on Environment and Development[1]

Because of their potential for self-reproduction, population and industrial capital are the driving forces behind exponential growth in the world system. Because of their potential for production, societies encourage their growth.

We assume in World3 that population and capital have the structural potential for both reproduction and production. We also assume that those potentials cannot be realized without continuous inflows of energy and materials and without continuous outputs of pollution and wastes.

People need food, water, air, and nutrients to grow, to maintain their bodies, and to produce new people. Machines need energy, water, and air plus an enormous variety of minerals, chemicals, and biological materials to produce goods and services, to maintain themselves, and to make more machines. According to the most fundamental laws of the planet, the materials and energy used by the population and the capital plant do not disappear. Materials are either recycled or they become wastes or pollutants. Energy is dissipated into unusable heat.

Figure 3-1 POPULATION AND CAPITAL IN THE GLOBAL ECOSYSTEM

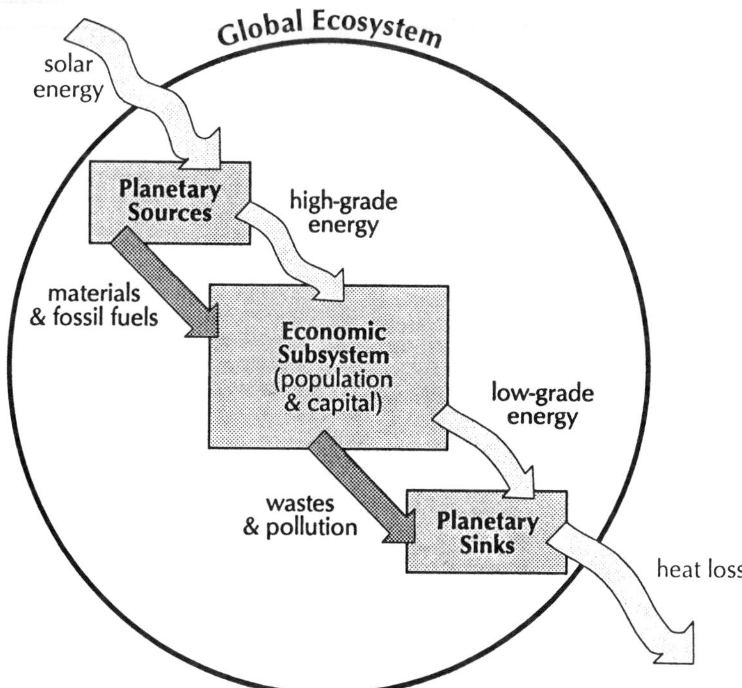

Population and capital are sustained by flows of fuels and nonrenewable resources from the planet, and they produce outflows of heat and waste, which contaminate the air, waters, and soils of the planet. (*Source*: *R. Goodland et al.*)

Population and capital draw materials and most forms of energy from the earth and return wastes and heat to the earth. There is a constant flow or *throughput* from the planetary *sources* of materials and energy, through the human economy, to the planetary *sinks* where wastes and pollutants end up (Figure 3-1). There are limits to the rates at which human population and capital can use materials and energy, and there are limits to the rates at which wastes can be emitted without harm to people, the economy, or the earth's processes of absorption, regeneration, and regulation.

Each resource used by the human economy—food, water, wood, iron, phosphorus, oil, and hundreds of others—is limited by both its sources and its sinks. The exact nature of these limits is complex, because sources and sinks are all part of a dynamic, interlinked, single system: the earth. Some limits are much more stringent than others. There are short-term limits (the amount of oil processed and waiting in storage tanks, for instance) and long-term limits (the amount of oil under the ground). Sources and sinks may interact, and the same natural feature of the earth may serve as both source and sink at the same time. A plot of soil, for example, may be a source for food crops and a sink for acid rain caused by air pollution. Its capacity to serve either of those functions may depend upon the extent to which it is serving the other.

World Bank economist Herman Daly has suggested three simple rules to help make order out of this complexity and to define the long-term or ultimately sustainable limits to throughput:

- For a *renewable resource*—soil, water, forest, fish—the sustainable rate of use can be no greater than the rate of regeneration. (Thus, for example, fish are harvested sustainably when they are caught at a rate that can be replaced by the remaining fish population.)
- For a *nonrenewable resource*—fossil fuel, high-grade mineral ore, fossil groundwater—the sustainable rate of use can be no greater than the rate at which a renewable resource, used sustainably, can be substituted for it. (For example, an oil deposit would be used sustainably if part of the profits from it were systematically invested in solar collectors or in tree planting, so that when the oil is gone, an equivalent stream of renewable energy is still available.)
- For a *pollutant* the sustainable rate of emission can be no greater than the rate at which that pollutant can be recycled, absorbed, or rendered harmless by the environment. (For example, sewage can be put into a stream or lake sustainably at the rate at which the natural ecosystem in the water can absorb its nutrients.) [2]

We will use these three criteria in this chapter to make a quick survey of the various forms of throughput and the states of their planetary sources and sinks. We will start with renewable resources and ask: Are they being used faster than they regenerate? Are their stocks falling? Then we will go on to nonrenewable resources, whose stocks by definition must be falling. For them we will ask: Are renewable substitutes being found? Will they be developed in time to support the functions of the human economy that are now dependent upon nonrenewables? Finally we will turn to pollutants and wastes and ask: Are they building up? Are their sinks overflowing or likely to overflow?

Those are questions to be answered not with the World3 model (nothing in this chapter depends upon that model) but with the global data, insofar as those data exist, source by source, sink by sink. For the moment we will ignore the interactions of one source or sink with another (for example, the fact that growing more food takes more energy, or that agricultural land usually expands at the expense of forest). We will need the computer model to keep track of such interactions, so we will come back to them in later chapters.

The limits we discuss here are the ones science happens to know most about. There is no guarantee that they are in fact the most limiting. The technologies we mention here are evolving. They will certainly be improved in the future. There will be surprises ahead, pleasant and unpleasant. But even given the incompleteness of human understanding about limits, we think the evidence presented in this chapter adds up to three clear points:

- Human society is now using resources and producing wastes at rates that are not sustainable.
- These excessive rates of throughput are not necessary. Technical, distributional, and institutional changes could decrease them greatly while maintaining and even improving the quality of life of the world's people.
- But even with much more efficient institutions and technologies, the limits of the earth's ability to support population and capital are close at hand, probably not more than a doubling or two away.

Renewable Sources

Food

Between 1950 and 1985 world grain production rose from around 600 million metric tons per year to over 1800 million metric tons per year. The average annual growth rate of grain production was 2.7%,

Figure 3-2 WORLD GRAIN PRODUCTION 1950–1990

Index (1950 = 100)

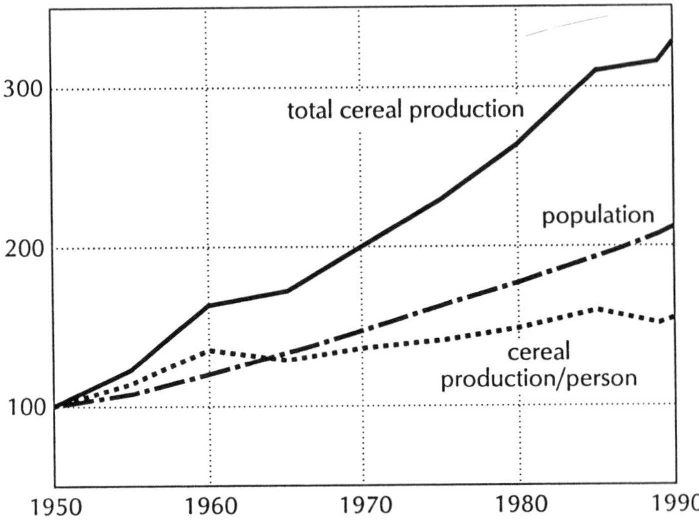

The world's farmers produced over three times as much grain in 1990 as they did in 1950. Because of population growth, however, per capita production in 1990 was only about 50% above the level at midcentury. (*Source: Food and Agriculture Organization.*)

slightly faster than the rate of population growth (Figure 3-2). The total amount of food produced in the world in 1989, if evenly distributed, could have fed 5.9 billion people a subsistence diet, 3.9 billion a moderate diet, or 2.9 billion a diet at the level of Europe. (The population that year was 5.2 billion.) These figures assume a 40% waste factor because of losses between harvest and consumption.[3]

The amount of food grown in an average year is sufficient to feed the present world population adequately, but not lavishly. Because of waste and unequal distribution, it feeds part of the population lavishly, part moderately, and another part totally inadequately.

Of the earth's more than 5 billion people over 1 billion at any one time are eating less food than their bodies require. Somewhere between 500 million and 1 billion people are chronically hungry. Each year 24 million infants are born underweight. In 1990 it was estimated

Figure 3-3 Grain Yields

Thousand kilograms per hectare per year

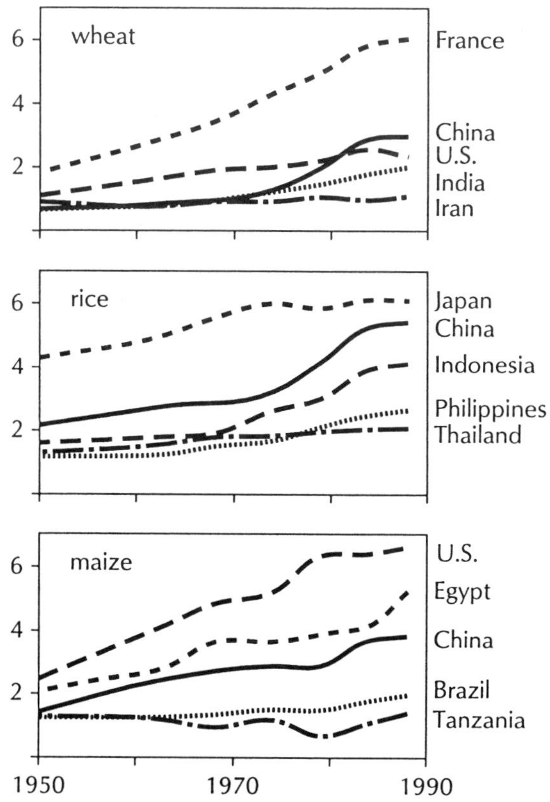

Yields of wheat, rice, and maize (corn) are high and beginning to level off in the industrialized world. In some industrializing nations, such as China, Egypt, and Indonesia, they are rising fast. In other less-industrialized nations they are still very low, with considerable potential for improvement. (In order to smooth out yearly weather variations, yields in these graphs have been averaged over 3-year intervals.) (*Source: Food and Agriculture Organization.*)

that 204 million children under the age of five were seriously undernourished.[4] Roughly 13 million people die every year of causes related to hunger. That comes to an average of 35,000 deaths from hunger every day. Most of those who die are children.

Hunger does not persist in the world because of physical limits—not yet, anyway. Food could be distributed more evenly, post-harvest losses could be reduced, and more food could be raised. For example, Figure 3-3 shows trends in grain yields in several countries. Yields in the highly industrialized countries are nearing their practical limits. But yields could be much higher in many industrializing countries.

In a thorough study of soils and climate in 117 countries of Latin America, Africa, and Asia, the U.N. Food and Agriculture Organization estimated that only 19 of these countries, with a combined population of 104 million, would *not* be able to feed their expected year-2000 populations from their own lands if they could use every hectare of useful land and get the highest yields technically possible. According to this study, if all cultivable land were allocated to food, if there were no loss to erosion, and if there were perfect weather, perfect management, and uninhibited use of agricultural inputs, the 117 countries studied could multiply their food output by a factor of 16.[5]

One obvious limit to food production is land.[6] Theoretical estimates of the potential amount of cultivable land on earth range from 2 to 4 billion hectares (depending upon what is considered "cultivable") of which roughly 1.5 billion are being cropped today. The area actually cultivated has declined slightly in the past twenty years, because land losses to erosion, salt buildup, urbanization, and desertification have somewhat exceeded the development of new agricultural land.[7]

As Figure 3-4 shows, under a wide variety of assumptions about the future the land limit is close, but extendable. The heavy solid line in Figure 3-4 shows the amount of land needed to maintain per capita food production at present levels, assuming the present world average of 0.28 hectares per person, and assuming two population futures. The lighter curves underneath show the land needed if average yields worldwide could be doubled or even quadrupled.

The shaded area in Figure 3-4 shows the amount of land possibly available. The upper edge of this area assumes the allocation of every feasible bit of land to food production and no further loss of cultivable land to urbanization or erosion. The lower edge assumes that the cultivated area will be maintained at its present value of 1.5 billion hectares. (If this area continues to be maintained by development of new land, leaving wasteland behind, it is in fact bringing the upper edge down, and cannot be sustained.)

Figure 3-4 Possible Land Futures

Billion hectares

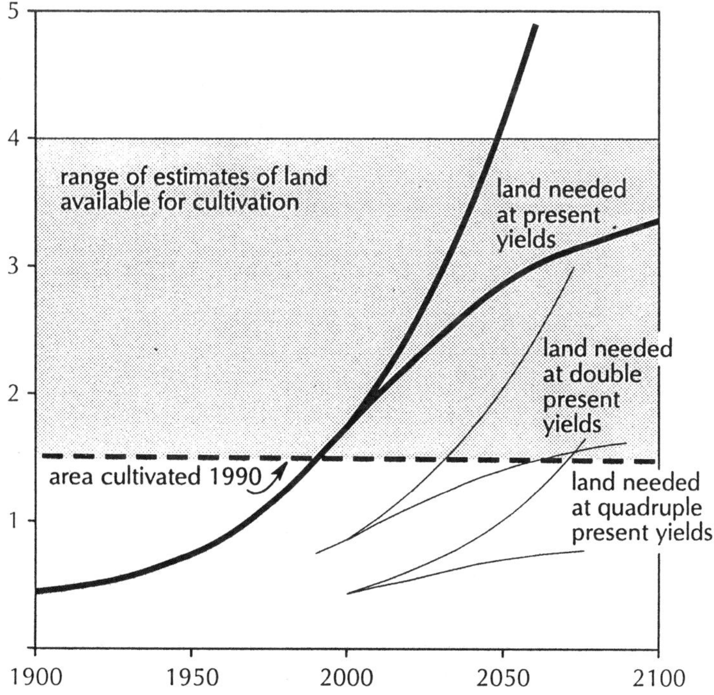

The heavy dashed and solid lines show extrapolations of potentially cultivable land and land required to maintain present per capita food production (if world population grows according to World Bank forecasts and according to continued exponential growth at present rates). The lighter lines show land requirement if crop yields are doubled and quadrupled. The shaded area shows the range of estimates of land that could hypothetically be brought into cultivation—much of which is presently forest land. (*Sources*: *G. M. Higgins et al.; World Resources Institute; R. A. Bulato et al.*)

You can see in Figure 3-4 how quickly exponential growth in population has moved the world from a situation of great land abundance to one of impending scarcity. There has been an overwhelming excess of potentially cultivable land throughout all human history, but within about thirty-five years (the last population doubling) there has arisen a sudden shortage.

But Figure 3-4 also shows how many response possibilities there might be, depending on the resilience of the resource base and the technical and social flexibility of humankind. If more land were developed or eroded land restored, if no more land were lost, if yields could double worldwide, every one of the present 5.4 billion people could have enough food, and so could the 12.5 billion that are projected to be on earth by the end of the next century. But if land erosion continues, if developing or restoring land proves too expensive, if another doubling of yield is too difficult or environmentally hazardous, if birth rates do not come down promptly the way the World Bank projects they will, food could become suddenly limiting not only locally, but globally.

Given the uncertainties illustrated in Figure 3-4, it is apparent that the world cannot afford to lose agricultural land. Yet land is being lost. Soil degradation is a notoriously difficult process to measure, but there is no doubt that it is widespread. When you look for quantitative estimates of its extent, you find partial but suggestive statements like these:

> During the 20 years since the first Earth Day in 1970, deserts expanded by some 120 million hectares, claiming more land than is currently planted to crops in China. . . . Over two decades . . . the world's farmers lost an estimated 480 billion tons of topsoil, roughly equivalent to the amount on India's cropland.[8]

> Short-sighted policies are leading to degradation of the agricultural resource base on almost every continent: soil erosion in North America; soil acidification in Europe; deforestation and desertification in Asia, Africa, and Latin America; and waste and pollution of water almost everywhere. . . . By the late 1970s soil erosion exceeded soil formation on about a third of U.S. cropland. . . . In Canada, soil degradation has been costing farmers $1 billion a year. . . . In India soil erosion affects 25–30 percent of the land under cultivation. Without conservation measures, the total area of rainfed cropland in . . . Asia, Africa, and Latin America will shrink by 544 million hectares over the long term.[9]

> In most areas of the Third World the land degradation problem is severe. It has been estimated that six to seven million hectares of agricultural land are made unproductive each year because of erosion. Waterlogging, salinization and alkalinization damage another

> 1.5 million ha. . . . The United Nations Environment Programme reported that in the early 1980s a total of 1,501 million hectares of rangeland and cropland in developing countries were undergoing at least moderate desertification.[10]

Loss of the agricultural resource base is a consequence of many factors, including poverty and desperation, expansion of human settlements, overgrazing and overcropping, mismanagement, ignorance, and economic rewards for short-term production rather than long-term stewardship.

There are other limits to food production besides land, among them the availability of water and the sinks for agricultural chemicals (which we will come to later in this chapter). Many parts of the world are already beyond some of these limits. Soils are eroding, irrigation is drawing down groundwater, runoff from agricultural fields is polluting surface and groundwater. These excursions beyond limits are not sustainable, and they are also not necessary.

Farming methods that conserve and enhance soils—such as terracing, contour plowing, composting, cover-cropping, polyculture, and crop rotation—have been known and used for centuries. Other methods particularly applicable in the tropics, such as alley cropping and agroforestry, are being demonstrated in experiment stations and on farms.[11] In both temperate and tropic zones high yields are being obtained sustainably without high rates of application of fertilizers and pesticides.[12] Millions of farmers in all parts of the world already follow soil-conserving and ecologically sound agricultural techniques. The challenge is to see that all farmers know and are able to practice these techniques. That is not a technical problem, it is a social one.[13]

If the flow of food through the human society were more efficient, less wasteful, and more evenly distributed, it would not be necessary to grow more. More food could be grown, however, and it could be done sustainably. But those are hypothetical statements. The present reality is that in many parts of the world the sources of food—land, soils, waters, soil nutrients—are falling and the sinks of pollutants from agriculture are overflowing. In those places the rates of agricultural throughput are already beyond sustainable limits. Unless rapid changes are made—changes that are entirely possible to make—the earth's exponentially growing population will have to continue to try to feed itself from a degrading agricultural resource base.

Water

At international meetings about resources we have frequently heard the statement that even in the 1990s some countries or regions will have to stop their growth or go to war, or both, because of shortages of water.

That statement comes from the intuition of hydrologists, an intuition informed by looking at graphs like the one in Figure 3-5. Figure 3-5 is only illustrative, because water is a regional, not a global, resource. Every regional water graph, however, has the same general characteristics as this global one—a limit, a number of factors that can expand that limit or that can make it unreachable, and exponential growth toward the limit.

In Figure 3-5 the physical limit is the total annual runoff of all the streams and rivers of the world. This is the renewable flow from which all freshwater inputs to the human economy are taken. It is a huge amount of water, 40,000 cubic kilometers per year. It would seem to be a far-off limit indeed, given current human water use of only 3500 cubic kilometers per year.[14]

In practice, however, the resource cannot be used to its full potential. Much of the runoff is seasonal. There is no way to store so much water. Therefore as much as 28,000 cubic kilometers per year flows to the sea in flood. That leaves only 12,000 cubic kilometers that can be contained and counted on as a sustained resource. Furthermore, some rivers flow where there are not many people, especially in the tropics and near the poles. The accessible stable runoff is therefore only about 7000 cubic kilometers per year.

However, Figure 3-5 also shows that human beings are raising the limit by building dams to trap floodwaters. By the end of the century human-built dams will increase the sustainable water supply by about 3000 cubic kilometers per year.[15] (Dams flood land, of course, and the river-basin land they flood is often prime agricultural soil.) There are other ways to raise the water limit, such as desalination of seawater, new settlements in uninhabited areas, and long-distance transport of

Figure 3-5 FRESH WATER RESOURCES

Cubic kilometers per year

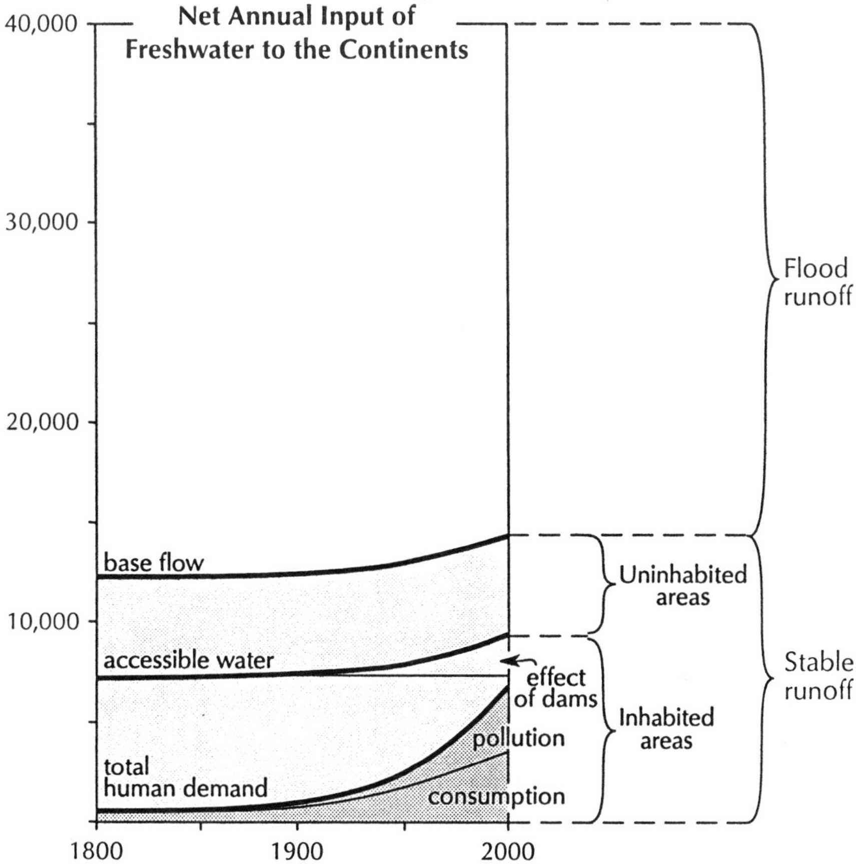

A graph of global fresh water resources and uses shows how quickly exponential growth in consumption and in pollution can approach the total amount of water that is stable and accessible. In 1950 human demand for fresh water was only about one-half the amount of water that was accessible. Only a strenuous dam construction program will leave some margin between demand and supply by the year 2000. (*Source*: *R. P. Ambroggi*.)

water. These changes can be important locally, but so far they are too small to show up on a global-scale graph.

Figure 3-5 also shows that there is monumental waste of the water resource. That is visible in the line showing pollution. The amount of water made unusable by pollution is almost as great as the amount actually used by the human economy. There is further waste, not shown in the figure, in the inefficient use of water.

Finally, there is exponential growth in demand, driven by both population and capital. Global water demand has been growing faster than the limit is being raised by dam building. And as the most favorable sites are taken, and as citizen opposition to dams increases, dam building will slow.

Globally water is in great excess, but because of operational limits and pollution, it can in fact support at most one more doubling of demand, which will occur in 20 to 30 years.[16] Even if it were possible to stop all pollution, trap every drop of flood, move either the water to the people or the people to the water, even if it were possible and desirable to capture the planet's full 40,000 cubic kilometers of annual runoff for human use, there would be enough water for only 3 to 4 more doublings–a mere 100 years away if current growth rates continue.

You don't have to wait for a global water shortage to see what happens when a society overshoots its water limit. You can look at the parts of the world that have already done it. What happens depends on whether the society is rich or poor, whether it has neighbors with water excess, and whether it gets along with those neighbors. Rich societies with willing neighbors, such as southern California, can construct canals, pipelines, and pumps to import water. Rich societies with vast oil reserves, like Saudi Arabia, can use fossil energy to desalinate sea water. Rich societies with neither, like Israel, can come up with ingenious technologies to use every drop of water with maximum efficiency and can shift their economies toward the least water-intensive activities. Societies with none of those options must develop severe rationing and regulation schemes. And poor societies experience famine and/or conflict over water.[17]

Most water-limited societies, rich and poor, give in to the temptation to draw down stocks of groundwater unsustainably. More than 4 million hectares of cropland in the United States are irrigated with water that is being pumped up faster than the aquifers can recharge. Buildings in Bangkok and Mexico City are sinking because the groundwater below them is being drawn away. Water levels in wells in Beijing are reported to be dropping at the rate of 1 meter per year; in Manila at the rate of 4 to 10 meters per year; in the Indian state of Tamil Nadu at the rate of 25 to 30 meters per year. Saltwater is intruding into falling freshwater aquifers in such coastal cities as Dakar, Jakarta, Lima, and Manila.[18]

Groundwater drawdown, water importation, and desalination are all strategies that can support locally, for a while, an economy that has grown beyond its water limit. None of these strategies can work globally or for very long. Water is not the most stringent limit everywhere. Where it is a stringent limit, it can be wasted less, polluted less, and managed better. But at some point those parts of the human economy that have already grown beyond their water limit will have to accept the simple fact that they cannot continue to grow exponentially against the renewable but fixed water budget of the earth.

Forests

Before humans invented agriculture there were 6 billion hectares of forest on Earth. Now there are 4 billion, only 1.5 billion of which are undisturbed primary forest.[19] Half of that forest loss has occurred between 1950 and 1990.

The United States (exclusive of Alaska) has lost one-third of its forest cover and 85% of its primary forest. Europe has essentially no primary forest left. Its remaining forests are managed plantations of just a few commercial tree species. China has lost three-fourths of its forests. The great remaining temperate forests are in Canada and Russia, where 1.4 billion hectares remain, half of them never harvested. Temperate-zone forests are now roughly stable in area, though many of them are declining in soil nutrients, species composition, wood quality, and growth rate.

The temperate-zone history of forest use will not be repeated in the tropics, because tropical soils, climates, and ecosystems are very different from temperate ones. Tropical forests are much richer in species, much faster-growing, but also much more vulnerable than temperate forests. There is no guarantee that they can survive even one extensive clear-cut without severe degradation of soil and ecosystem integrity. Experiments are currently underway to find a method of logging tropical forests selectively or in strips to allow regeneration. But most of the large-scale logging now taking place is essentially treating the tropical forest not as a renewable resource, but as a nonrenewable one.

In the tropics half the original forest cover is gone. Half of what remains has been logged and degraded. No one is sure exactly how fast the forest is being cleared. The first authoritative attempt to assess tropical deforestation rates, conducted by the FAO in 1980, came up with a figure of 11.4 million hectares lost per year. By the mid-1980s that rate had climbed to over 20 million hectares per year. After some policy changes, particularly in Brazil, the rate of loss by 1990 had apparently come down to around 17 million hectares per year.

Figure 3-6 shows the extent and speed of forest loss in one small country, Costa Rica, whose recent history demonstrates some of the worst and best forestry policies in the world. Much of the forest was cleared in Costa Rica in order to expand cattle ranching for beef export. Many of the new pastures proved unsustainable. Within a few years they were grazed down, eroded, and abandoned. On steep hillsides and in heavy rains there were landslides, which destroyed roads and villages. Silt from eroded lands filled up reservoirs behind hydropower dams or washed into the oceans, where it buried and killed coral reefs and destroyed fisheries. The land will bear scars for a long time from Costa Rica's few decades of intensive beef production.

Costa Rica has acted belatedly but effectively to preserve its remaining forests and even to restore some that have been lost. Nearly all the remaining primary forests are in national parks or other protected areas. Through a series of debt-for-nature exchanges, Costa Rica is finding the resources to put into place the infrastructure and expertise that will maintain these protected forests for scientific study and ecotourism, which may create more employment and international exchange, sustainable over the long term, than was ever possible from the ill-fated cattle pastures.

The reasons for forest clearing vary from one tropical country to

Figure 3-6 FOREST COVER IN COSTA RICA 1940–1984

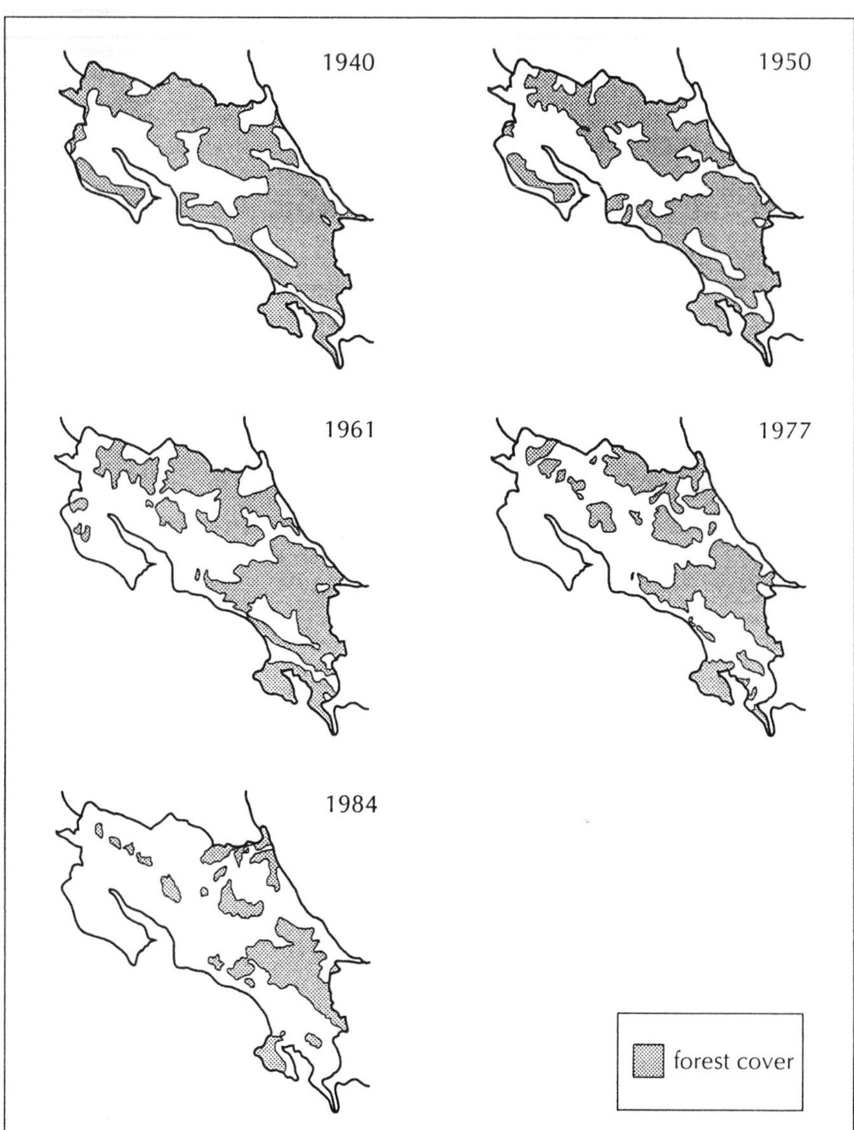

The forested area of Costa Rica has been greatly diminished over a period of just 50 years. Most of the remaining forests, however, have now been protected. (*Source*: C. *Quesada*.)

another. The perpetrators include multinational timber and paper companies; governments anxious to increase exports and pay external debts; rich local landowners, ranchers, and farmers; and poor local people scrambling for firewood or a patch of land on which to grow food. These actors often work in concert, the government inviting the companies in, the companies harvesting and taking away the wood, and the poor moving in along the logging roads to find land for settlement.

What is the future of the tropical forest? One can use several different assumptions to extrapolate forward the current rates of loss. Figure 3-7 illustrates some possibilities. The remaining primary tropical forest in 1990 covered approximately 800 million hectares (about 330 million of which were in Brazil). Roughly 17 million hectares, or 2.1%, were cut that year.

- If the clearing rate increases exponentially, say at the rate the population of the tropical countries is growing (about 2.3% per year), the forest will be gone in 30 years. This curve depends on an assumption that the driving forces behind forest loss will grow exponentially.
- If the clearing rate stays constant at 17 million hectares per year, the forest will be gone in 47 years. This possibility is shown by the straight line in Figure 3-7. It is based on the assumption that the forces that cause forest destruction will neither strengthen nor weaken.
- If the clearing rate remains a constant percentage of the remaining forest (2.1% per year), the area cut will be slightly less each year than the year before. The forest area will decline gradually toward zero; most of it will be gone after 100 years. This projection assumes that each cut makes the next cut less likely, perhaps because the nearest, most valuable forests are taken first.

The real future will be more complex than any of these theoretical curves, as populations grow, as increasing remoteness and declining quality make logging more difficult, and as political pressures increase to protect at least some of the remaining forest. The point is simply that as world demand for construction timber, paper products, and fuelwood is growing (Figure 3-8), forests in nearly every part of the world are disappearing.

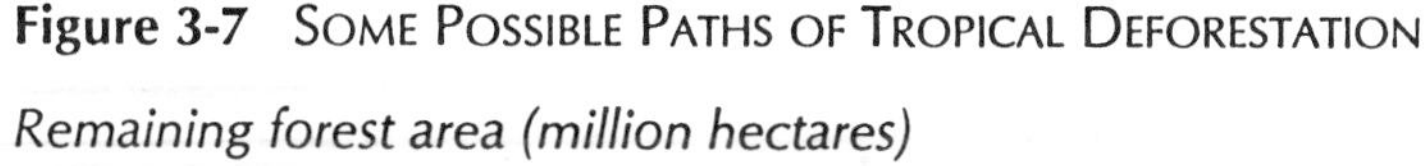

Figure 3-7 SOME POSSIBLE PATHS OF TROPICAL DEFORESTATION

Remaining forest area (million hectares)

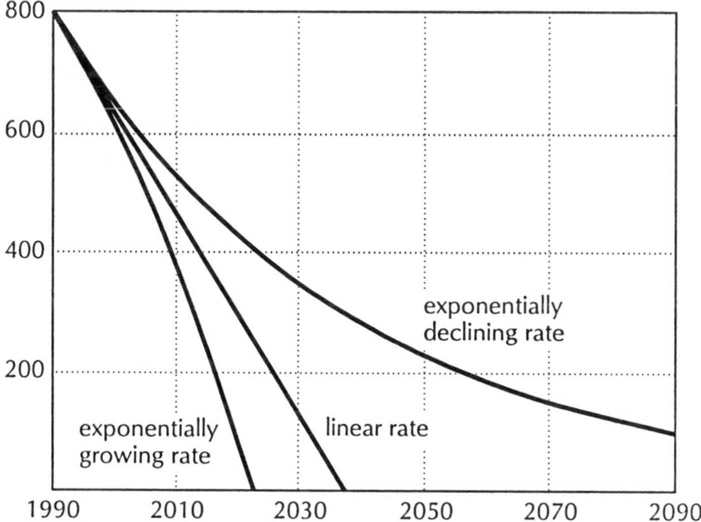

Estimates of the future loss of tropical forests depend upon assumptions about demographic, legal, and economic trends. The three scenarios, shown in this plot, indicate a range of possible futures if there is no concerted and effective agreement to protect the remaining forests. If the initial loss of 17 million hectares per year increases with population at about 2.3% per year, the forest will be gone by 2020. If the rate of loss is constant at 17 million hectares per year, the forest will be gone by about 2040. If the rate of loss is 2.1% of the remaining area each year, the forest will decline gradually toward zero over 100 years or more.

Forest cutting in China exceeds regrowth by 100 million cubic meters a year. India's forests shrink by 1.5 million hectares per year; its demand for wood outstrips its estimated annual growth by a factor of 7. Logging in Canada's province of British Columbia in 1989 was 30% higher than sustainable yield. Softwood harvests on the West Coast of the United States during the 1980s exceeded sustainable yield by 25% on industry-owned land and by 61% in government-owned national forests. Firewood scarcity is critical in India and much of sub-Saharan Africa. The World Bank forecasts that over the next decade the number of tropical countries exporting wood will drop from 33 to 10.[20]

Figure 3-8 World Roundwood Production

Billion cubic meters of wood per year

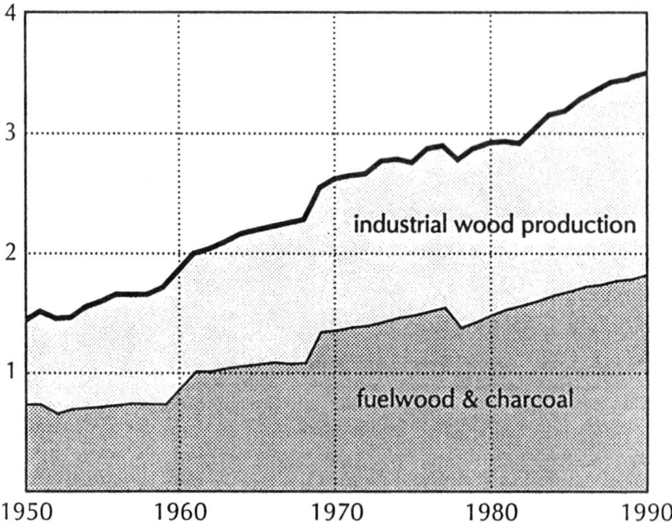

Roundwood production equals total volume of wood removed from forests and from trees outside forests. Commodities made from roundwood include sawlogs, veneer logs, pulpwood, other industrial roundwood, and fuelwood. Statistics include recorded volumes, as well as estimated unrecorded volumes. (*Source: Food and Agriculture Organization.*)

Overharvesting is one threat to forests. Another is pollution. Three-fourths of European forests, long managed for sustained yield, are damaged by air pollution and acid rain. Pollution damage to forests is estimated to be costing Europe at least $30 billion per year, about as much as the value of the output of West Germany's iron and steel industries, and three times as much as Europe's annual expenditures on air pollution control. Even if Europe cut its emissions of sulfur oxides, nitrogen oxides, and ammonia by 60% to 80%, some of its forests would still be at risk.[21] The enormous forest of European Russia—equal in size to all the other forests of Europe—is threatened by both air pollution and by harvesting well above sustainable levels.[22]

The loss of forests is a problem for more reasons than just the loss of forest-based products. A standing forest is a resource in itself, performing vital functions that are beyond economic measure. Forests create soil, moderate climate, control floods, and store water against drought. They cushion the erosive effects of rainfall, hold soil on slopes, and keep rivers and seacoasts free from silt. They harbor and support most of the earth's species of life. The tropical forests alone, which cover only 7% of the earth's surface are believed to be the home of at least 50% of the earth's species. Forests take in and hold a great stock of carbon, which helps balance the stock of carbon dioxide in the atmosphere and thus combats the greenhouse effect (more on that later in this chapter).

As with soil, as with water, the present unsustainable rates of forest use are not really necessary. Harvest rates could be greatly reduced by eliminating waste and increasing recycling. The United States has the world's highest rate of paper use per capita (317 kg per person per year), half of which goes into quickly discarded packaging, and only 29% of which is recycled. Japan recycles 50% of its paper, but it also uses high-quality tropical hardwoods to make plywood panels for molding concrete, which are thrown away after one or two uses. Half of U.S. wood consumption could be saved by increasing the efficiency of sawmills, plywood mills, and construction, by doubling paper recycling, and by reducing the use of disposable paper products. Similar steps taken throughout the industrialized countries could be combined with fuel-efficient stoves throughout the industrializing countries, to reduce the world's demand for wood.[23]

Logging, especially in tropical forests, could be conducted in such a way as to reduce its negative impact on soils, streams, and unharvested trees. High-yield forest plantations could be expanded greatly, not at the expense of primary forests, but on already logged or marginal lands. High-yield agriculture could reduce the need for growing populations to move into forests. Direct and indirect government subsidies to logging industries could be removed, so that the prices of wood products signal more realistically their actual cost.

These are all measures to reduce throughputs of forest products and to bring forest harvest rates back down below sustainable limits.

None of these measures is impossible. Every one of them is being practiced somewhere in the world, but not in the world as a whole. And so the forests continue to shrink.

The Other Species of Life

There may be anywhere from 10 million to 100 million species of life on earth. Only 1.4 million have been named and classified by humans. Since no one knows within a full order of magnitude how many species there are, no one can know how many are being lost. But the number of extinctions is almost certainly increasing exponentially. That is deducible from the rate at which habitat is disappearing. For example:

- Madagascar is a biotic treasure house; its eastern forest houses 12,000 known plant species and 190,000 known animal species, at least 60% of which are found nowhere else on earth. More than 90% of that forest has been eliminated. Scientists estimate that at least half the original species have gone with it.
- Western Ecuador once contained 8000 to 10,000 plant species, about half of them endemic. Each species of plant supports between 10 and 30 animal species. Since 1960 nearly all the western Ecuadorian forests have been turned into banana plantations, oil wells, and human settlements. The number of species lost in just 25 years is estimated to be 50,000.[24]

Most extinctions are happening, as you might expect, where the most species are: in tropical forests, coral reefs, and wetlands. Wetlands are probably even more endangered than tropical forests. Like tropical forests they are places of intense biological activity, including the breeding of many species of fish. Only 6% of the earth's surface is wetland—or was. It is estimated that about half the world's wetlands have been lost to dredging, filling, draining, and ditching. That doesn't count what might be degraded by pollution.

Estimates of global extinction rates start with measures of habitat loss, which are fairly accurate. They go on to assume how many species might be in the habitat that is lost, and those assumptions are uncertain by a factor of 10. Then they assume a relationship between habitat loss and species loss. The rule of thumb is that 50% of the species will remain even if 90% of the habitat is gone.

These calculations are subject to considerable argument.[25] But, as with other numbers we've been trying to grapple with in this chapter, their general direction is clear. No one doubts that species are disappearing at an accelerating rate. Estimates range from 10 to over 100 species lost every day. Ecologists say there has not been such a wave of extinctions on earth since the events that eliminated the dinosaurs at the end of the Cretaceous Age 65 million years ago.

Species loss is one way of measuring the human impact on the biosphere. Another way of measuring was undertaken by ecologists at Stanford University a few years ago. They calculated how much of the biological activity of the planet is appropriated for the use of human beings. Their results are astonishing. They found that humans commandeer 25% of the photosynthetic product of the earth as a whole (land and sea), and 40% of the photosynthetic product on land![26]

That figure takes a little explanation. Ecologists define the *net primary production* (NPP) of the biosphere as the amount of energy captured from sunlight by green plants and fixed into living tissue. The NPP is the base of all food chains. Every other living thing eats plants, or eats some other creature that eats plants, or eats a creature that eats a creature that eats plants, and so on. The NPP, therefore, is the energy flow that powers all nature.

Humans consume *directly* only about 3% of the land-based NPP through food, animal feed, and firewood. *Indirectly* another 36% of NPP (on land) goes to crop wastes, forest burning and clearing, desert creation, and conversion of natural areas to settlements. That calculation does not include reduction of primary production by pollution—that effect is not yet calculable on a global scale. Humans *control* about 40% of the NPP on land; humans may *affect* much more than that through pollution.

If the 40% figure is even approximately correct, it poses some interesting questions about the next doubling of human population and economic activity, only 20 to 30 years away. What would the world be like if human beings co-opted 80% of the NPP? Or 100%?

No one is sure. Some ecologists say that a world where human beings use 100% of the NPP would look like the Netherlands or England—no real wilderness, the landscape under human control, many wild species extinguished, not much room for expansion or mistakes, but a livable world.

Others point out that the Netherlands and England import food, feed, wood, and fiber and therefore depend on far more than 100% of the NPP of their own territory.[27] Some countries can do that; the world as a whole cannot. A world at the NPP limit might look like the Sahel, some ecologists say, or like China.

One thing is certain. As humans take more of the primary productivity of the earth for themselves and the life forms of their choice (such as corn and cows), they leave less for other life forms. The result is a loss of economic value: game, fish, chemicals, medicines, foods may be disappearing with species that no one has even identified. There is also a spiritual and esthetic loss, a loss of colorful companions in creation. There may be, for all anyone knows, a loss of critical pieces that hold together ecosystems. There is certainly a loss of genetic information that has taken billions of years to evolve—and that humanity is just beginning to learn how to read and use.

Somewhere along the path of NPP usurpation, there lie limits. Long before the ultimate limits are reached, the human race becomes economically, scientifically, aesthetically, and morally impoverished.

Nonrenewable Sources

Fossil Fuels

The human economy's energy throughput grew between 1860 and 1985 by a factor of 60. World energy consumption has continued to climb, unevenly but inexorably, through wars, recessions, price instabilities, and technical changes (Figure 3-9). Most of that energy flows through the industrialized world. The average European uses 10 to 30 times as much commercial energy[28] as the average person in a developing country, and the average North American uses 40 times as much.[29]

The World Energy Conference projected in 1989 that a business-as-usual continuation of population and capital growth would increase world energy demand by another 75% by the year 2020 and that the

Figure 3-9 World Energy Use

Millions of terajoules per year

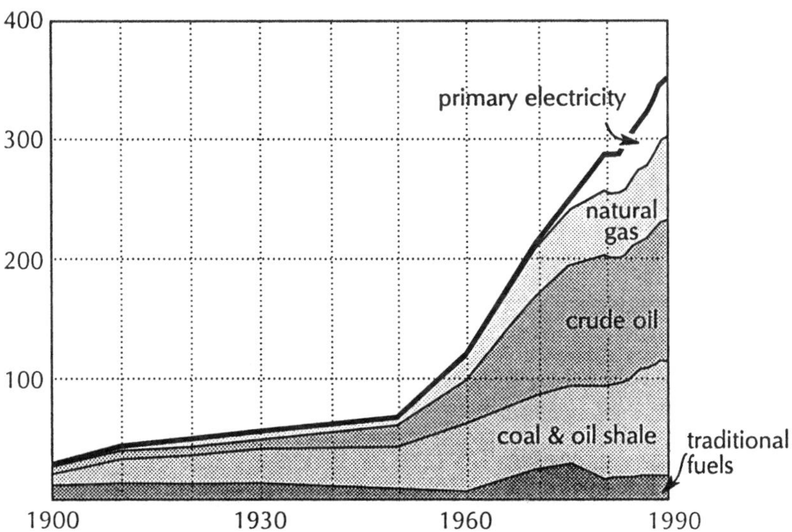

Rates of energy use and the relative contributions of different sources reflect the evolution of technology as well as the growth of human population. Although fossil fuels still dominate the primary energy supply, coal's share peaked around 1920, when it provided more than 70% of all the fuel consumed; oil's share peaked in the early 1970s at slightly more than 40%. Natural gas, which is less polluting than either oil or coal is expected to contribute more in the future to global energy use. Primary electricity on this graph includes both hydroelectric power and nuclear power. (*Sources: United Nations; G. R. Davis.*)

mix of fuels supplying this energy would continue to be dominated by the nonrenewable fossil fuels: coal, oil, and gas.[30] At present 88% of the commercial energy used in the world comes from fossil fuels.

Between 1970 and 1990 the world economy burned 450 billion barrels of oil, 90 billion tons of coal, and 1100 trillion cubic meters of natural gas. Over that same twenty-year period, however, new deposits of oil, coal, and gas were discovered (and some old ones were reappraised upward). Therefore, although fossil fuel consumption rates are now higher than they were in 1970, as shown in Table 3-1, the ratio of

Table 3-1 ANNUAL PRODUCTION AND RESERVE/PRODUCTION RATIOS FOR OIL, COAL, AND GAS, 1970 AND 1989

Fuel	*1970 production (per year)*	*1970 R/P (years)*	*1989 production (per year)*	*1989 R/P (years)*
Oil	16.7 billion barrels	31	21.4 billion barrels	41
Coal	2.2 billion tons	2300	5.2 billion tons	326 (hard coal) 434 (soft coal)
Gas	30 trillion cu. ft.	38	68 trillion cu. ft.	60

known reserves to production (R/P, or the number of years known resources will last if production continues at its current rate) has gone up for both oil and gas. The apparent drop in R/P for coal comes from incomparable methods of estimation; as the table indicates, coal is by far the most abundant fossil fuel.

Do these rising reserve/production rations mean that there were more fossil fuels to power the human economy in 1990 than there were in 1970?

No, of course not. There were 450 *fewer* billion barrels of oil, 90 billion *fewer* tons of coal, and 1100 trillion *fewer* cubic meters of natural gas. Fossil fuels are nonrenewable resources. When they are burned they turn into carbon dioxide, water vapor, sulfur dioxide, and a number of other combustion products, which do not, on any time scale of interest to humanity, come back together to form fossil fuels again. Rather, they are wastes and pollutants that enter planetary sinks.

Those who see the discoveries of the past twenty years as proof that there are no limits to fossil fuels are looking at only part of the energy system:

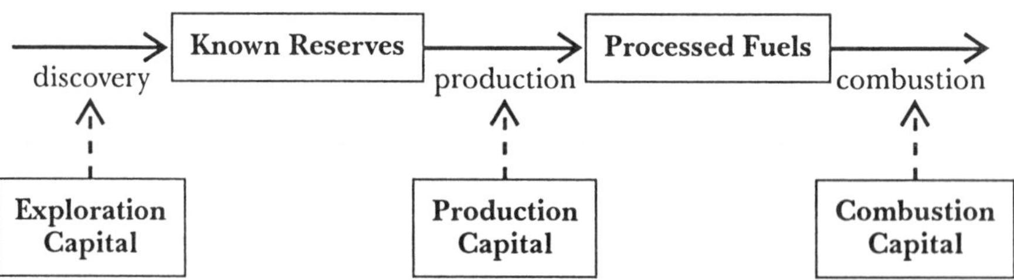

The process of *discovery* uses exploration capital (drilling rigs, airplanes, satellites, a sophisticated array of sounders and probes) to find fossil fuel deposits in the earth and thereby to increase the reserves that have been identified but not yet extracted. The process of *production* pulls that stock out of the ground, using mining, pumping, transport, and refining capital, and delivers it to the places where processed fuels are stored. Then the *combustion* capital (furnaces, automobiles, electricity generators) burns the fuels.[32]

As long as the rate of discovery exceeds the rate of production, the stock of known reserves goes up. But the diagram above shows just the *economic* part of the system, the part people happen to watch and measure. A more complete diagram would include the ultimate sources and sinks for fossil fuels:

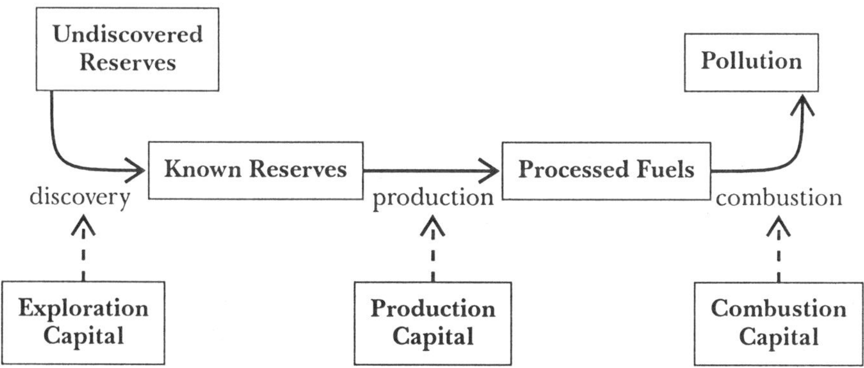

As production reduces the stock of known reserves, energy companies invest in discovery to replenish it. But every discovery comes from the ultimate stock of fossil fuels in the earth, which is not replenished. The stock of undiscovered reserves may be very large, but it is finite and nonrenewable and declining.

At the other end of the flow, combustion produces pollutants, which enter the ultimate sink—the biogeochemical processes of the planet, which recycle pollutants, or render them harmless, or are poisoned by them. Pollutants of various types are also emitted at every other stage of the fossil-fuel flow, from discovery through production, refinement, transportation, and storage. (A major source of groundwater pollution in the United States, for example, is leaking underground oil storage tanks.)

The flow of fossil fuels is limited by both sources and sinks, but for different fuels the source and sink limits differ greatly in their severity. There is such an enormous amount of coal that its use will most likely be limited by sinks, especially by the already overflowing atmospheric sink for carbon dioxide. The limits to oil are becoming apparent at both ends. Its combustion produces greenhouse gases and other pollutants, and it will certainly be the first fossil fuel to be depleted at the source.

Estimates of undiscovered oil reserves range greatly and can never be certain, but in some parts of the world depletion is already critical to economics and politics (see Table 3-2). In 1988, 24% of world oil production came from the Middle East and 21% from the USSR; these two regions have between them 72% of known oil reserves and 40% of estimated undiscovered reserves.

Oil depletion will not appear as a complete stop, a sudden drying-up of the spigot. Rather, it will show up as lower and lower returns to exploration effort, increasing concentration of the remaining reserves in the Middle East, and finally a peak and gradual decline in total world production. The United States provides a case study in depletion. Its enormous original oil endowment is more than half gone. Its domestic oil production peaked in the late 1960s, and its oil demand must be met more and more by imports (see Figure 3-10).

Of all the fossil fuels, natural gas emits the least pollution per energy unit, and therefore it may rapidly replace oil and coal in the future—which will speed up its source depletion to an extent that will surprise those who don't fully appreciate the dynamics of exponential growth. Figures 3-11 and 3-12 show why.

In 1989 the world reserve/production ratio for natural gas was 60 years, which means that if current known reserves continued to be used at 1989 consumption rates, they would last until the year 2050. Two things will happen to make that simple extrapolation wrong. One is that more reserves will be discovered. The other is that future use will not be constant.

Table 3-2 WORLD OIL RESERVES AND PRODUCTION, SELECTED REGIONS AND NATIONS

Region or nation	*Cumulative production to 1988*	*1988 production*	*Known reserves*	*Estimated undiscovered reserves*
		Billion barrels		
World	610.1	21.3	922.1	275–945
Middle East	160.2	5.1	584.8	66–199
USSR	103.6	4.5	80.0	46–187
U.S.	152.7	3.0	48.5	33–70
Asia & Pacific	36.8	2.2	42.8	37–148
Africa	46.4	2.0	58.7	20–92
South America	57.9	1.4	43.8	18–86
Western Europe	15.7	1.4	26.9	11–56
Mexico	15.7	0.9	27.4	15–75
Canada	14.3	0.5	7.0	9–57
Eastern Europe	6.8	0.1	2.0	1–4

Suppose, for purposes of illustration, that enough recoverable gas will eventually be discovered to use at the 1990 world rate not for 60 but for 240 years. (That is a generous estimate. The general consensus is that yet-undiscovered reserves will be roughly the same size as current proved reserves, and there is a systematic tendency for fossil fuel resource estimates to overshoot the actual amounts finally obtainable.[33]) If the 1990 use rate remained constant, gas reserves would go down linearly, as illustrated by the diagonal line in Figure 3-11 and would last 240 years. But if consumption continues to grow at the rate at which it has grown over the past twenty years, about 3.5% per year, the 240-year reserve would plummet exponentially as shown by one of the lines in Figure 3-11. It would be exhausted not in 2230, but in 2054; it would last not 240 but only 64 years.

If, to reduce some forms of pollution and escape oil depletion, the world calls upon natural gas to carry the energy load now handled by coal and oil, the growth rate could well be faster than 3.5%. If it were 5% per year, the "240-year supply" would be exhausted in 50 years.

Figure 3-12 shows how discoveries would have to increase in order

Figure 3-10 U.S. OIL PRODUCTION AND EXPLORAT'ON HISTORY

Billion barrels per year

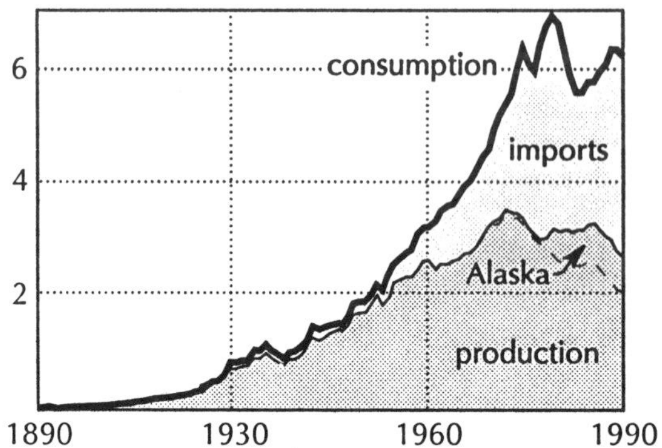

Barrels per foot drilled in lower 48 states

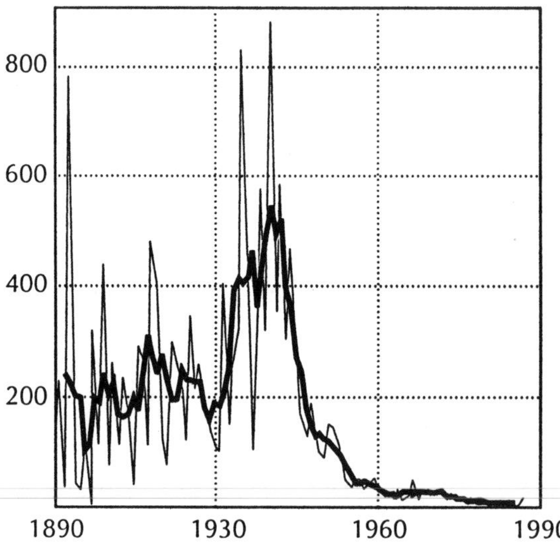

Depletion of United States oil reserves was first signaled by the drop in yield to exploration after 1940. U.S. domestic oil production peaked in 1970, and production in the lower 48 states has since dropped by 40%. Even new discoveries in Alaska have not allowed production to recover to its 1970 level. (*Sources: American Petroleum Institute; C. J. Cleveland et al.*)

Figure 3-11 Depletion of the World's Gas Reserves Assuming Different Rates of Growth in Consumption

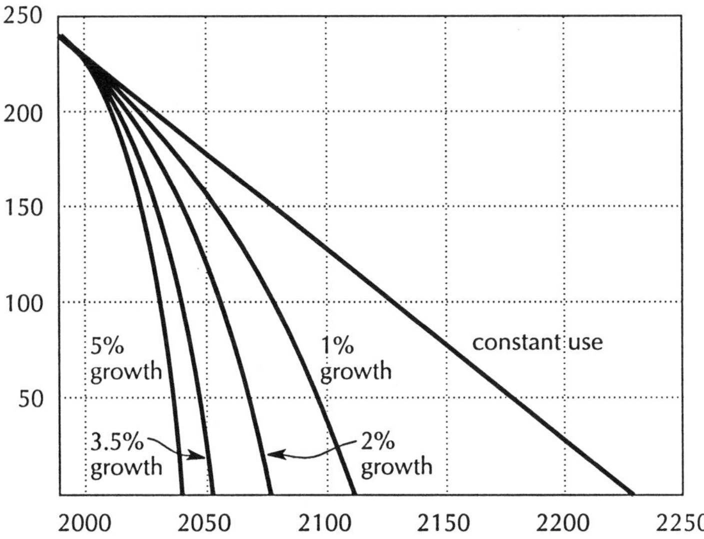

If discoveries eventually quadruple the present global reserves of natural gas, the current consumption rate of the fuel can be sustained until 2230. But depletion of oil combined with environmental problems of coal could shift reliance to gas. If gas consumption were to continue to grow at its present rate of 3.5% per year, an amount of gas equal to 4 times the currently known reserves would be consumed by 2054.

to permit a steady growth of natural gas consumption at 3.5% per year. By the mathematics of exponential growth, the amount of gas discovered would have to double every 20 years. Every two decades as much gas would have to be discovered as had been discovered in all previous history.

The point is not that the world is about to run out of natural gas. The considerable reserves that remain will be essential as a transition fuel on the way to more sustainable energy sources. The point is that fossil fuels are not sustainable, are surprisingly limited, especially when used exponentially, and should not be wasted. On the scale of human

Figure 3-12 NECESSARY GAS DISCOVERIES TO MAINTAIN A 3.5% PER YEAR GROWTH RATE

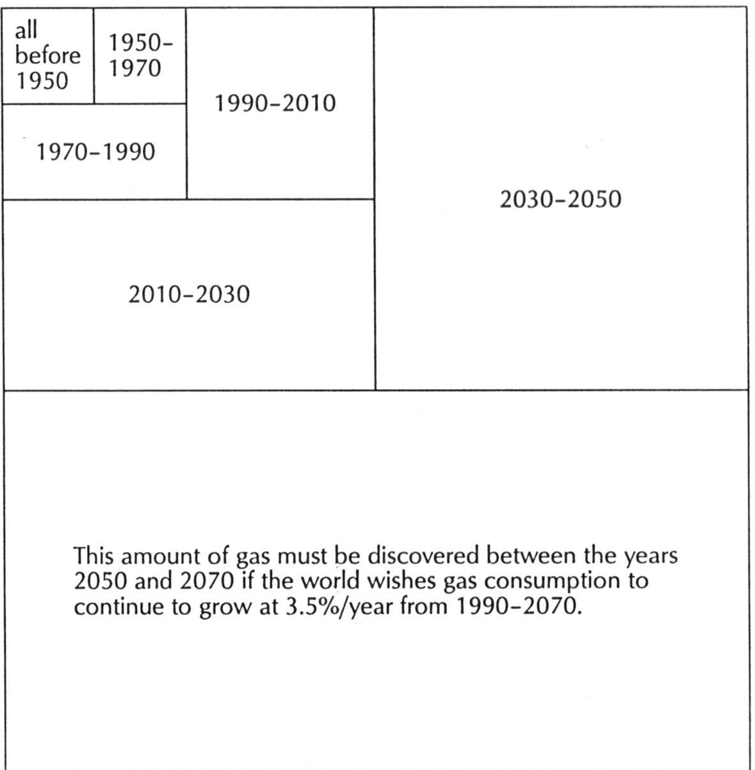

If the rate of growth of natural gas consumption continues at 3.5% per year, that means that every 20 years an amount of new gas must be discovered that is equal to all the previous discoveries of history. (*Source*: *A. A. Bartlett.*)

history, the era of fossil fuels will be a short blip, either because of their source limits or their sink limits. A society that expands its fossil-fuel burning capital without planning ahead to renewable substitutes is likely to find itself very suddenly beyond its energy limits.

There are renewable substitutes for fossil fuels. There need not ever be global energy scarcity. Two energy options are available that are sustainable from the source, environmentally supportable, technically feasible, and increasingly economic. One of them, greater *efficiency*, can be implemented quickly. The other, *solar-based renewables*, will take only a little longer.

Energy efficiency means producing the same final energy services—lighted, heated, and cooled rooms, transport for people and freight, pumped water, turning motors—but using less energy to do so. It means the same or better material quality of life, usually at less cost—not only less direct energy cost, but also less pollution, less drawdown of domestic energy sources, less conflict over siting facilities, and, for many countries, less foreign debt and less military cost to maintain access to or control over foreign resources.

Efficiency technologies, from better insulation to smarter motors, are improving so quickly that estimates of the energy needed to accomplish any given task have to be revised downward every year. A compact fluorescent light bulb will give the same amount of light as an incandescent one but use only one-fourth as much electricity. Insulating superwindows in all U.S. buildings could save twice as much energy as the nation now gets from Alaskan oil. At least ten automobile companies have built prototype cars that drive 30 to 60 kilometers on a liter of gas (65 to 130 miles per gallon), and leading-edge technical discussions are now beginning to speak of 70 km/liter (160 mi/gallon) vehicles. Contrary to popular belief, these efficient cars pass all tests for safety, and some cost no more to build than current models.[34]

Calculations of how much energy could be saved through efficiency depend on the technical and political biases of the people who do the calculating. On the conservative end of the range, it seems certain that the North American economy could do everything it now does, with currently available technologies and at current or lower costs, using half as much energy. That would bring North America to the present efficiency levels of Western Europe and Japan—and it would reduce the worldwide drain on oil by 14%, coal by 10%, gas by 15%. Similar or greater efficiency improvements are possible in Eastern Europe and the less-industrialized world.

The optimists say that's only the beginning. They believe that Western Europe and Japan, already the most energy-efficient economies of the world, could increase their efficiencies by factors of 2 to 4 with technologies already available or easily foreseeable within twenty years. Some calculations suggest that with improved efficiency the world as a whole could keep its total energy throughput at or below the current level with no reduction in productivity, comfort, or convenience in the rich countries, and with steady economic growth in the poor countries.[35]

Efficiency of that magnitude would make it possible to supply most or all of the world's energy from *solar-based renewable sources*—sun, wind, hydropower, and biomass such as wood, corn, or sugarcane. The sun pours much more energy upon the earth every day than human beings can ever use. Total human use of fossil fuel constitutes a flow of power equal to about 5 trillion watts (terawatts). The constant inflow of the sun to the earth's surface is 80,000 terawatts.

Technical advances in capturing the sun's energy have been slower than those in raising efficiency, but they have been steady nonetheless. In 1970 photovoltaic (PV) electricity was generated at a capital cost of $150 per watt. By 1990 the cost had dropped to $4.50 per watt.[36] A cost reduction by another factor of 3 to 4 will make PV electricity competitive with large-scale coal-fired plants, even without environmental costs being counted in the price of power. In less-industrialized countries PV is already the most cost-effective choice for villages and irrigation projects that cannot afford the capital cost of connecting to a distant electric grid.

Solar thermal and wind-powered electricity in appropriate locations are already cost competitive (Figure 3-13) and more such technologies are on the horizon.[37] Studies for the U.S. Department of Energy say that within forty years the United States could get 57% to 70% of the total energy it uses now from sun, wind, water, geothermal, and biomass.[38] Since at least half the energy the country uses now could be saved by higher efficiency, that means a totally renewable energy future could be possible.

Renewable energy sources are not environmentally harmless and they are not unlimited. Windmills require land and access roads. Some kinds of solar cells contain toxic materials. Hydroelectric dams flood

Figure 3-13 Costs of Electricity from Solar Energy and Wind Power

Cents per kilowatt hour

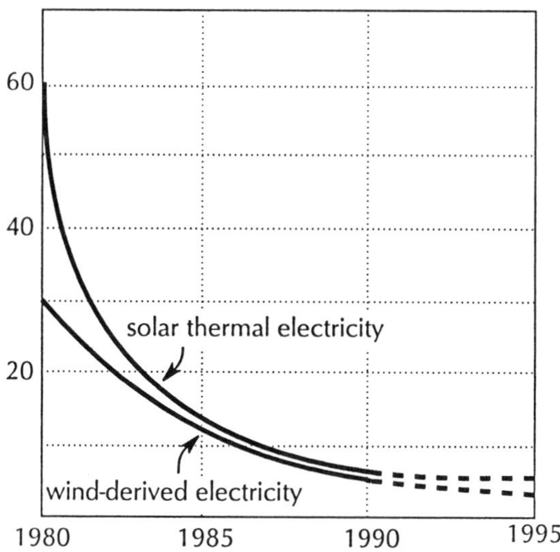

Between 1980 and 1990 the cost of electricity generated by solar thermal facilities and by windmills has fallen by more than a factor of 5. Both technologies are now competitive with conventional electricity generating technologies. (*Source:* G. *Heaton et al.*)

land and ruin free-flowing streams. Biomass energy is only as sustainable as the agriculture or forestry practices that produce the biomass. Some solar sources are dilute and intermittent and require large collection areas and complex storage mechanisms,[39] and all require physical capital and careful management. And renewable energy sources are rate-limited; they can flow forever, but only at a fixed pace. They cannot support any size population and capital plant growing at any rate. But they can provide the energy base for the sustainable society of the future. They are abundant, widespread, and varied. Their associated pollution flows are lower and generally less harmful than those of fossil or nuclear energy.

There is no scarcity of energy on earth. If the most sustainable, least polluting sources are used with high efficiency, it should be not only possible but affordable to power the needs of the human race sustainably. That is not the direction energy policy is taking in most countries now. But it is a feasible and beneficial direction for any country that wants to bring itself below its throughput limits.

Materials

Only 8% of the world's people own a car. Many nations plan to double or triple their highways, schools, and hospitals. Hundreds of millions of people—no one knows exactly how many—live in inadequate houses or have no shelter at all, much less lights, refrigerators, or television sets. If there are going to be more people in the world, and if they are to have more or better housing, health services, education, cars, refrigerators, televisions, they will need steel, concrete, copper, aluminum, plastic, and many other materials.

One sometimes hears of a "post-industrial" society that will use fewer materials, because the economy will consist less of industry and more of services. That idea does not take into account the extent to which services depend on an industrial base and on materials brought from all over the world. Amory Lovins once wrote about a common piece of service-sector machinery, a typewriter (his description would also apply to computers, laser printers, and fax machines):

> The typewriter I am now using probably contains Jamaican or Surinam aluminum, Swedish iron, Czech magnesium, Gabonese manganese, Rhodesian chromium, Soviet vanadium, Peruvian zinc, New Caledonian nickel, Chilean copper, Malaysian tin, Nigerian columbium, Zairean cobalt, Yugoslav lead, Canadian molybdenum, French arsenic, Brazilian tantalum, South African antimony, Mexican silver, and traces of other well-travelled metals. The enamel may contain Norwegian titanium; the plastic is made of Middle Eastern oil (cracked with American rare-earth catalysts) and of chlorine (extracted with Spanish mercury); the foundry sand came from an Australian beach; the machine tools used Chinese tungsten; the coal came from the Ruhr; the end product consumes, some might say, too many Scandinavian spruces.[40]

This description is useful not only to point out the intertwined routes along which the industrial economy moves materials, but also to emphasize that every piece of that typewriter originated from the earth. When the typewriter's useful lifetime is over, it will most likely end up in the earth again.

The stream of materials from the earth through the economy and back to the earth can be diagrammed in the same way as the flow of fossil fuels, with one exception. Unlike fossil fuels, materials such as metals, concrete, plastic, and glass do not turn into combustion gases after use. They either accumulate somewhere as solid waste, or they are reclaimed and recycled, or they are broken down, pulverized, leached, vaporized, or otherwise dispersed into soils, waters, or the air.

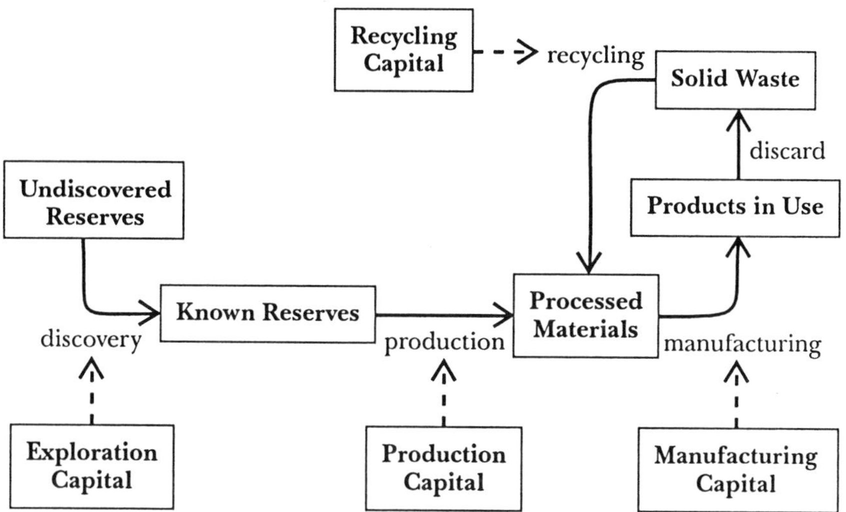

Figure 3-14 shows the history of per capita materials use in a country supposedly moving into the post-industrial mode—the United States. It illustrates two points: first that materials use rises and falls with economic cycles, and second that materials use per person does level off. There is a limit to the amounts of steel, concrete, and copper that even wealthy people can put through their lives.

That limit is high, however, at least if the American lifestyle is to be any model for the rest of the world. For most metals, the average use rate of a person in the industrialized world is 8 to 10 times the use rate of people in the nonindustrialized world. If an eventual 12.5 billion people all consumed materials at the rate of the average late-twentieth-century American, that would require an increase in worldwide steel production by a factor of 7, copper by a factor of 11, and aluminum by a factor of 12.

Most people have an intuitive sense that such materials flows are neither possible nor necessary. They are not possible because of the limits to the earth's sources and sinks, and because all along the way from source to sink the processing, fabricating, handling, and use of materials leaves trails of pollution. They are not necessary because the per-person material throughputs of the rich nations of the late twentieth century, like their food, water, wood, and energy throughputs, are wasteful. A good life could be supported with much less tearing-up of the planet.

There are signs that the world is learning that lesson. Figure 3-15 shows the world's consumption of metals from 1930 to 1988. Something happened in the mid-1970s to interrupt what had been a smooth exponential growth trend. There are several theories about what did happen, all of them probably right:

- The oil price shocks in 1973 and again in 1979 made the prices of energy-intensive metals rise sharply, and therefore some people could not afford metal-intensive products.
- The same higher prices, plus environmental laws and solid waste disposal problems, encouraged materials recycling.
- Those pressures brought about a technical revolution; plastics, ceramics, and other materials were substituted for metals. Products from automobiles to soft-drink cans were made lighter.
- The economic downturn of the early 1980s reduced materials demand.
- During that downturn the heavy manufacturing sectors were most depressed, so basic metal demands were reduced disproportionately.[41]

Figure 3-14 TRENDS IN U.S. CONSUMPTION OF MATERIALS

Kilograms per person per year

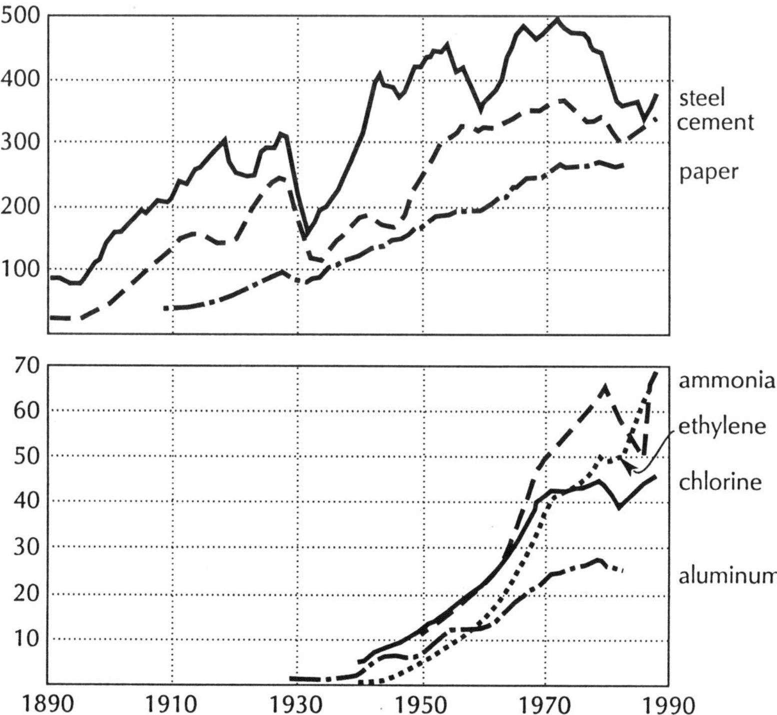

Trends in per capita consumption of seven materials exemplify the overall use of basic materials in the United States. Among the seven are three "traditional" materials: steel, cement, and paper. The others are "modern" materials: aluminum, ammonia, chlorine, and ethylene. Per capita use of the traditional materials has leveled off, except for the ups and downs of economic cycles. Per capita use of the newer materials rose continuously until the economic downturn of the early 1980s. (*Sources: E. Larson et al.; U.S. Bureau of the Census; United Nations.*)

The economic reasons for the slower growth in materials consumption may be temporary; the technical changes will probably be permanent, as will the environmental pressures to reduce material flows. Poor communities have always reclaimed and reused materials because of scarce sources. Rich communities are relearning how to recycle be-

Figure 3-15 WORLD METAL CONSUMPTION

Billion metric tons per year

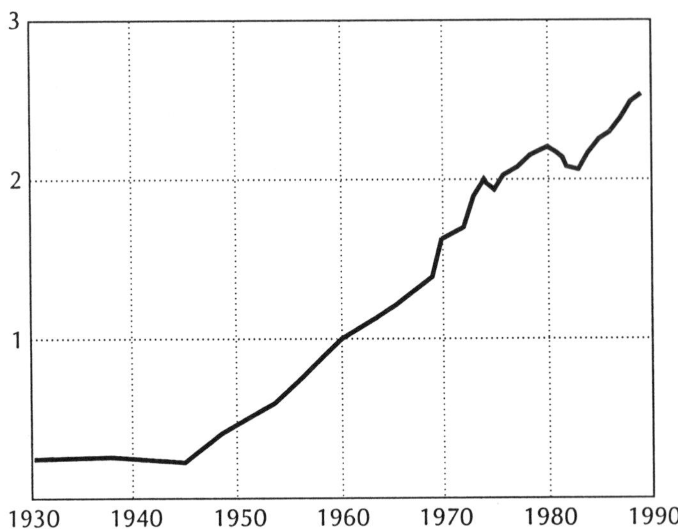

Total world metal consumption showed a slight downturn in the economic recession of the early 1980s, but then continued to rise. Since the 1970s, growth has been more linear than exponential. (*Sources: E. N. Cameron; U.S. Bureau of Mines.*)

cause of scarce sinks. In the process they are turning recycling from a labor-intensive to a capital- and energy-intensive activity. They use mechanized compost turners, shredders and screening systems, digesters, and sludge mixers, and they hire management companies to set up waste recovery programs for industries or municipalities.

Forward-looking manufacturers are designing products from teapots to cars with final disassembly and recycling in mind. A new BMW car has a plastic body designed for easy recycling. Plastics are now marked with their resin type, and fewer types are mixed together, so they can be separated and reused.

Separating and recycling materials after use is a step toward sustainability. It begins to move materials through the human economy the way they move through nature—in cycles. In nature the waste from one process becomes an input to another process. Whole sectors of ecosystems, particularly in the soils, work to take nature's waste materials apart, separate them into usable pieces, and send them back into living creatures again. The modern human economy is finally developing a recycling sector too.

But recycling trash is only dealing with the final and least problematic end of the materials stream. A rule of thumb says that every ton of garbage at the consumer end of the stream has also required the production of 5 tons of waste at the manufacturing stage and 20 tons of waste at the site of initial resource extraction (mining, pumping, logging, farming).[42] The best ways to reduce these flows of waste are to increase the useful lifetimes of products and to reduce material flows at their source.

Increasing product lifetime through better design, repair, and reuse (as, for example, in using returnable bottles and in washing cups instead of using throwaways) is more effective than recycling, because it doesn't require crushing, grinding, melting, purifying, and refabricating recycled materials. Doubling the average lifetime of any product will halve the energy consumption, the waste and pollution, and the ultimate depletion of all the materials used to make it.

Source reduction means finding a way of performing the same job with less material. It is the equivalent of energy efficiency and, like energy efficiency, its possibilities are enormous. In 1970 a typical American car weighed 4000 pounds, nearly all of it metal. Now the average new car weighs 2400 pounds, 180 pounds of which are plastic. Computer circuits are carried on minute silicon chips instead of heavy ferromagnetic cores. A small compact disk holds as much music as two large vinyl records. One hair-thin strand of ultra-pure glass can carry as many telephone conversations as 625 copper wires and with better sound quality.

Instead of the high temperatures, severe pressures, harsh chemicals, and brute force that have characterized manufacturing processes since the beginning of the industrial revolution, scientists are beginning to understand how to use the intelligence of molecular machines and of genetic programming. Breakthroughs in nanotechnology and biotechnology are beginning to allow industry to carry out chemical reactions the way nature does, by careful fitting of molecule to molecule.[43]

The possibilities for recycling, greater efficiency, increased product lifetime, and source reduction in the world of materials are exciting. On a global scale, however, they have not yet reduced the vast material flow through the economy. At best they have slowed its rate of growth. And billions of people still want cars and refrigerators, and those billions are growing exponentially. Though most people in the 1990s are more aware of sink limits than of source limits for material throughputs, continued growth in materials demand will eventually run into source limits as well. Many of the materials most useful to human society occur only rarely in concentrated form in the earth's crust, and they are being depleted just as fossil fuels are being depleted.

Geologist Earl Cook once illustrated with a table how unusually concentrated, and how rare, most mineable ores are.[44] He compared the "cutoff grade" for common metal ores with their concentration in ordinary rock. The cutoff grade is the lowest concentration of ore that is economically usable. For any mineral the cutoff grade may go down if capital, energy, and technology allow leaner ores to be processed or if price rises. But Table 3-3 shows that the cutoff grade would have to go down by factors of several hundred or several thousand before plain rock would become mineable for most minerals. The energy necessary to process such dilute mineral concentrations would never be affordable, much less environmentally supportable.

The table shows that iron and aluminum and possibly titanium are truly abundant in the earth's crust. From the source end those three metals can be considered essentially unlimited. The other minerals are, like the fossil fuels, scarce and precious, formed by geological processes of tremendous force over millions of years, nonrenewable, and steadily depleting.

Figure 3-16 shows what mineral depletion looks like–gradually decreasing ore concentration. Figure 3-17 shows the consequence of depletion. As the amount of usable metal in the ore falls below 1%, the amount of rock that must be mined, ground up, and treated per ton of product rises with astonishing speed. As the average grade of copper

Table 3-3 ECONOMICALLY IMPORTANT MINERALS: RATIOS OF MINEABLE CUTOFF GRADE TO AVERAGE CRUSTAL ABUNDANCE

Element	*Average crustal abundance (%)*	*Mineable cutoff grade (%)*	*Ratio*
Mercury	0.0000089	0.1	11,200
Tungsten	0.00011	0.45	4,000
Lead	0.0012	4	3,300
Chromium	0.011	23	2,100
Tin	0.00017	0.35	2,000
Silver	0.0000075	0.01	1,330
Gold	0.00000035	0.00035	1,000
Molybdenum	0.00013	0.1	770
Zinc	0.0094	3.5	370
Manganese	0.13	25	190
Nickel	0.0089	0.9	100
Cobalt	0.0025	0.2	80
Phosphorus	0.12	8.8	70
Copper	0.0063	0.35	56
Titanium	0.64	10	16
Iron	5.820	20	3.4
Aluminum	8.3	18.5	2.2

ore mined in Butte, Montana, fell from 30% to 0.5% the tailings produced per ton of copper rose from 3 tons to 200 tons. This rising curve of waste is closely paralleled by a rising curve of energy required to produce each ton of final material. Metal ore depletion hastens the rate of fossil fuel depletion.

The world economy uses about 2 billion tonnes of nonfuel minerals per year. That high rate of material flow reduces ore grades, increases energy use and waste production, fills up dumps, and emits pollutants all along the way. Even if there were no further growth, present rates of material use would be unsustainable in the long term. If a growing world population is to live in material sufficiency in the future, all the source-reduction and recycling technologies now on the horizon will be urgently needed. Materials will be treated as the limited, and precious gifts from the earth they actually are. The idea of a "throwaway society" will become obsolete.

Figure 3-16 The Declining Quality of Copper Ore Mined in the United States, 1906–1990

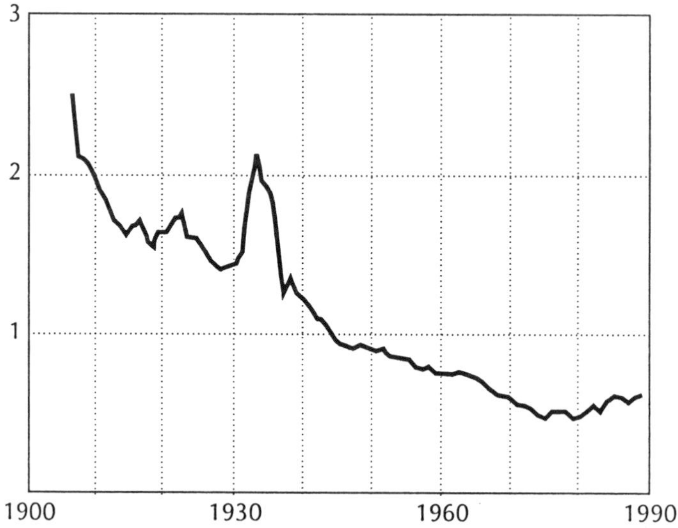

Ores averaging between 2% and 2.5% copper were mined in the United States before 1910. Since then, there has been a persistent decline in average grade. The peak in the 1930s and the slight rise in the 1980s were caused by economic downturns that shut down marginal mines and left functioning only those with the richest ores. (*Source*: *U.S. Bureau of Mines.*)

Sinks for Pollution and Wastes

At the time of the 1972 Stockholm Conference on the Environment there were no more than ten nations with environmental ministries or agencies. Now there are well over a hundred. The record of these new environmental protection institutions during their first twenty years has been mixed. It would be too simple to conclude that the world has solved its pollution problems—or that there has been no progress at all.

The greatest successes have come with specific toxins that are harmful to human health and that could be singled out and simply banned. Figure 3-18 shows, for instance, that since the use of lead in

Figure 3-17 DEPLETION OF MINERAL ORES GREATLY INCREASES THE MINING WASTES GENERATED IN THEIR PRODUCTION

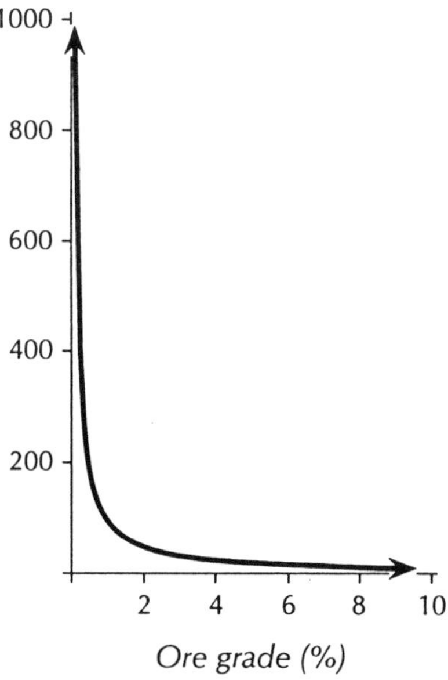

As the average grade of ore declines through depletion from 8% or more to 3%, there is a barely perceptible increase in the amount of mining waste generated per ton of final metal. Below 3% wastes per ton increase dramatically. Eventually the cost of dealing with the wastes will exceed the value of the metal produced.

gasoline and the pesticides DDT and dieldrin were banned in Belgium, Japan, and the Netherlands respectively, their levels in the environment and in human tissue have gone down. Populations of birds that were experiencing reproductive failure because of DDT have begun to recover.

In the industrialized countries after determined effort and considerable expense there has been partial success in decreasing some, but not all, of the most common air and water pollutants. Figure 3-19 shows that in the G7 nations[45] sulfur dioxide emissions have been cut by almost 40% by scrubbers on smokestacks and shifts to low-sulfur fuels. The pollutants carbon dioxide and nitrogen oxide are chemically difficult to scrub; they have been held roughly constant for twenty years, in spite of economic growth, mainly because of gains in energy efficiency.

The history of pollutants in the Rhine River provides an illustration of the triumphs and disappointments of water pollution control. Oxygen levels in the Rhine have been greatly improved (Figure 3-20), primarily because of investments in sewage treatment systems. The toxic heavy metal cadmium has stopped rising exponentially because of increasingly strict regulations against dumping into the Rhine, but it now permeates bottom sediments and does not break down chemically; therefore it remains at high levels. Chloride levels remain high because of political difficulties; the downstream nations have not found a way of applying effective pressure against the main chloride sources, which are salt mines in Alsace. Nitrogen comes from fertilizer drainage from agricultural lands. Its sources are too dispersed to be gathered into a sewage treatment system. The only way it can be reduced is through changed farming practices throughout the Rhine watershed.

The industrialized nations have managed to decrease greatly some of the most visible and easily handled pollutants (such as black, smoky particulates). They have held the line on others in spite of considerable growth in emission sources. The United States, for example, has spent $100 billion on wastewater treatment facilities over twenty years. It has halved the amount of organic pollution per volume of municipal effluent, but the amount of municipal effluent has doubled, so water quality in many places has remained roughly the same.[46] In the past twenty years California has reduced pollution emissions per car by 80% to 90%. Over the same period, however, the number of cars has risen 50% and the number of miles driven per car has climbed 65%.[47]

The Netherlands National Institute of Public Health and Environmental Protection calculates that to do better than just holding the line, to stop serious damage to soils and waters in the Netherlands, sulfur dioxide emissions will have to be reduced a *further* 90%, nitrogen oxides 70%, ammonia 80%, and phosphorus 75%.[48] The Institute says, "If the technological options are increasingly used up within the available time span, a fundamental reorientation of our expectations about the nature and size of 'economic' growth appears unavoidable."[49]

Figure 3-18 DECREASING HUMAN CONTAMINATION BY DDT, DIELDRIN, AND LEAD IN THREE COUNTRIES

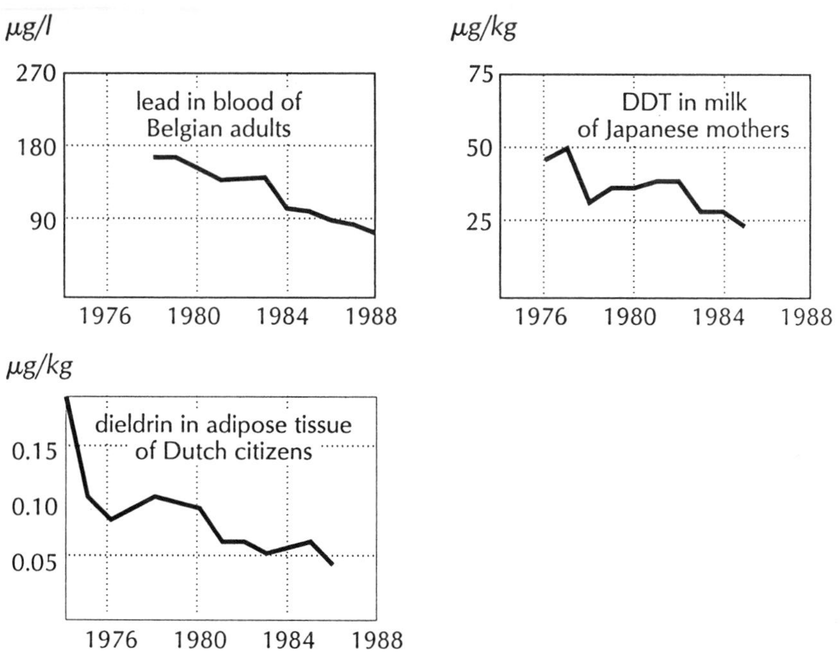

The most dramatic environmental improvements have come from outright bans on toxic substances such as lead in gasoline (in Belgium) and the pesticides DDT (in Japan) and dieldrin (in the Netherlands). (*Sources: United Nations Environmental Programme; G. Ducoffre et al.*)

That's the record in rich countries with money to spend on pollution abatement. The worst air and water pollution levels in the world are now found in Eastern Europe and the Third World, where billion-dollar abatement efforts are simply unimaginable. And that's the record for the kinds of pollutants that are chemically and politically the easiest to abate.

The most intractable pollutants, so far at least, are nuclear wastes, hazardous wastes, and wastes that threaten global biogeochemical processes, such as the greenhouse gases. They are chemically the hardest to sequester or detoxify, physiologically the hardest for the senses to detect, and economically and politically the most difficult to regulate.

Figure 3-19 TRENDS IN EMISSIONS OF SELECTED AIR POLLUTANTS

(1970 = 100)

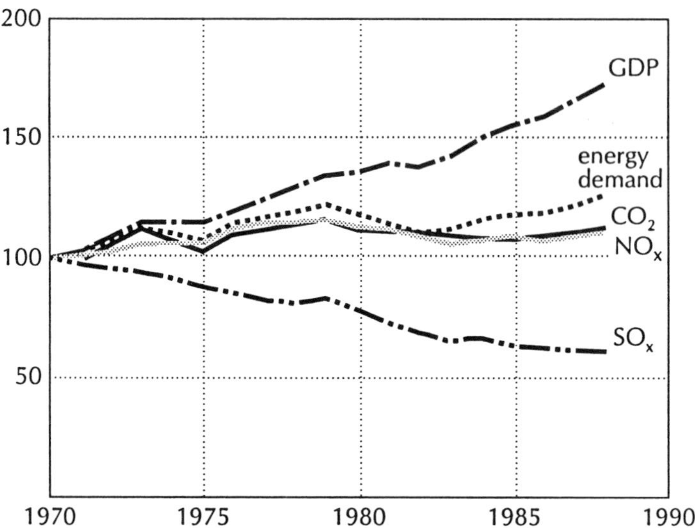

The G7 countries have made significant efforts to achieve energy efficiency and emission controls. Although their economies have grown by almost 60% since 1970, their emissions of CO_2 and NO_x have remained almost constant (mainly because of energy efficiency) and their emissions of sulfur oxides have decreased by 40% (because of both energy efficiency and active abatement technologies). (*Source*: *OECD*.)

No nation has solved the problem of nuclear wastes. They are hazardous to all forms of life, both by outright toxicity and mutagenicity. Nature has no way of rendering them harmless. They disintegrate by their own inner timetable, which for some can be decades, centuries, or even millennia. As byproducts of nuclear power production they are accumulating exponentially, stored underground or in water pools within the containment vessels of nuclear reactors, in hopes that someday the technical and institutional creativity of humankind will come up with some place to put them.

The most intractable hazardous wastes are human-synthesized chemicals. Since they have never before existed on the planet, no organisms have evolved to break them down and render them harmless.

Figure 3-20 RHINE RIVER POLLUTION

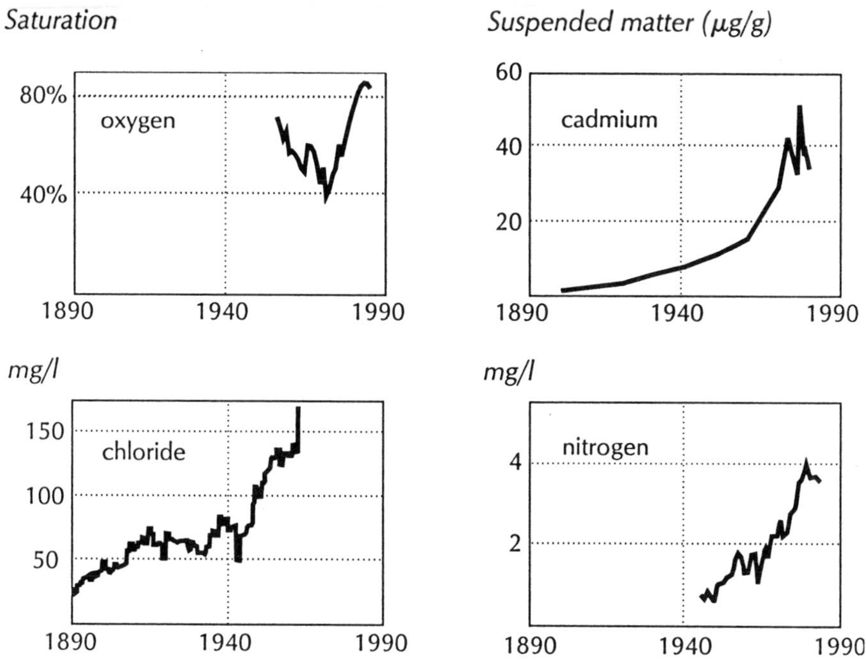

Data on the Rhine River show some successes and some failures in pollution control. Sewage treatment plants have increased oxygen levels to nearly normal. Cadmium levels have been reduced by about a quarter since their peak in the late 1970s, but they are still very high. Chloride from mining wastes and nitrogen from fertilizer runoff have not improved significantly. (*Sources*: *K. Mall; World Resources Institute; I. F. Langeweg.*)

A total of 65,000 industrial chemicals are now in regular commercial use. Toxicology data are available on fewer than 1% of them. Every day 3 to 5 new chemicals enter the marketplace. Eighty percent of these chemicals are not tested for toxicity.[50] Every day one million tons of hazardous wastes are generated in the world, 90% of them in the industrialized world. On an average day in the United States there are five industrial accidents involving hazardous waste.[51] Norway has recently discovered 7000 sites where the soil is contaminated with hazardous chemicals and metals; the government estimates that cleaning them up will cost $3 to $6 billion.[52] As the industrialized countries begin to find soils and groundwaters poisoned by decades of irresponsible chemical dumping, there is increasing pressure on the industrializing nations of the South to accept manufacturing processes or waste dumps that are no longer considered tolerable in the North.

And then there are the contaminants that pollute the great energy and materials flows of the earth as a whole. These global pollutants, no matter who generates them, affect everyone. The most dramatic example of global pollution has been the effect of the industrial chemicals called chlorofluorocarbons on the stratospheric ozone layer. The ozone story is a fascinating one, because it illustrates humankind's first unambiguous confrontation with a global limit. We think it is so important that we tell it fully in Chapter 5.

Many scientists believe that the next global limit humanity will have to deal with is the one called the greenhouse effect, or the heat trap, or global climate change.

Scientists have known for over a hundred years that carbon dioxide traps heat and increases the temperature of the earth, like a blanket, or more accurately like a greenhouse that lets the sun's energy in but hinders it from going back out. This "greenhouse effect" is a natural phenomenon and a beneficent one, which warms the earth and makes it habitable. But too much warming, caused by human-generated carbon dioxide from fossil-fuel burning and by deforestation, could cause a global climate change. Over the past twenty years it has been discovered that other "greenhouse gases" emitted by human activity are also building up exponentially in the atmosphere: methane, nitrous oxide, and the same chlorofluorocarbons that are threatening the ozone layer (Figure 3-21).

Global climate change cannot be detected quickly, because the weather from day to day or year to year is naturally variable. Climate is the long-term average of weather; therefore it can only be measured over decades. Evidence for warming, however, is beginning to accumulate. The eight hottest years (globally averaged) in the past century were, in increasing order: 1980, 1989, 1981, 1983, 1987, 1988, 1991, 1990 (Figure 3-22). Long-term studies of Canadian lakes are showing an increased ice-free season of three weeks, which is changing the relative populations of aquatic species. Caribbean corals are turning white and dying, because, some biologists believe, sea temperature is rising. Satellites show a shrinking snow cover over the Northern Hemisphere.[53]

None of these observations *proves* that the earth is warming because of atmospheric increase in greenhouse gases. Even if warming is occurring, the meaning of global climate change for future human activity or ecosystem health is not known for sure. Some politicians have escalated that uncertainty to a state of high confusion. Therefore it is important to state clearly what is known with certainty:[54]

- It is certain that human activities, especially fossil-fuel burning and deforestation, are increasing the atmospheric concentration of greenhouses gases. These gases have been monitored for decades. Their historical concentrations can be measured from bubbles of air caught in layers of ice drilled from the polar icecaps. There is no doubt whatsoever about their increase.
- The greenhouse gases trap heat that otherwise would escape from the earth into space. That is a well-known consequence of their molecular structure and spectroscopic absorption frequencies.
- Trapped heat will increase the temperature of the earth over what it would otherwise be.
- The warming will be unequally distributed, more near the poles than near the equator. Because the earth's weather and climate are largely driven by temperature differences between the poles and the equator, winds, rains, and ocean currents will shift in strength and direction.
- On a warmer earth the ocean will expand and sea levels will rise. If the warming is sufficient to melt polar ice in large quantities, sea levels will rise significantly.

There are three large uncertainties. One is what the global temperature would otherwise be without human interference. If long-term climatological factors unrelated to the increase in greenhouse gases happen to be cooling the planet, then the greenhouse gases will counteract those factors, but the counteracting trends might not add up to a

Figure 3-21 GLOBAL GREENHOUSE GAS CONCENTRATIONS

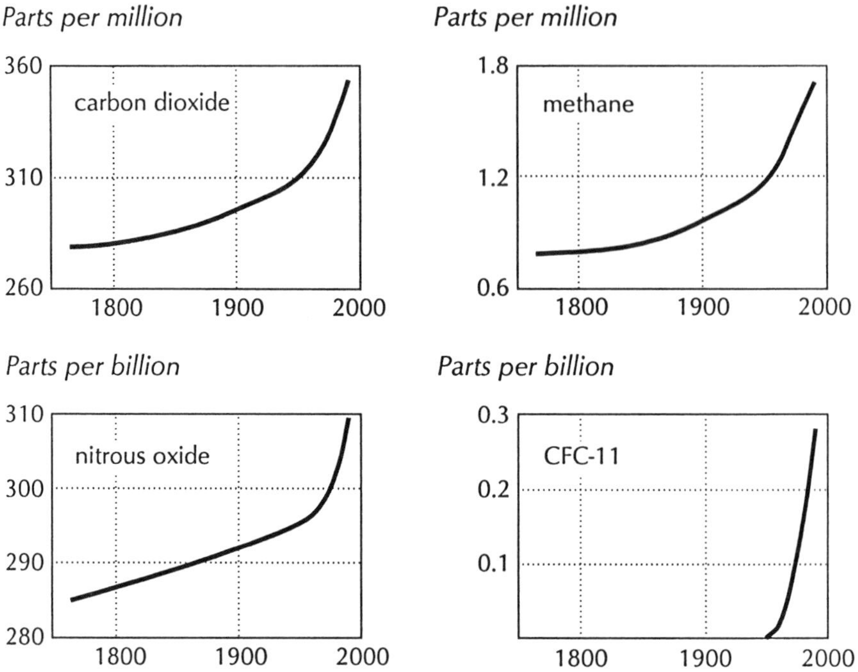

Carbon dioxide, methane, nitrous oxide, and chlorofluorocarbons (CFCs) all reduce emissions of heat from the earth to outer space. Hence all serve to increase the temperature of the earth. The atmospheric concentration of these gases, except the CFCs which were only recently synthesized, has been increasing since before 1800. (*Source: World Meteorological Organization.*)

net warming. A second uncertainty is what, exactly, a warming planet would mean for temperatures, winds, currents, precipitation, ecosystems, and the human economy in each specific place on Earth.

The third uncertainty has to do with feedbacks. Carbon flows and energy flows on Earth are immensely complex. There may be self-corrective mechanisms, negative feedback processes, that will stabilize the greenhouse gases or the temperature. One of them is already operating: the oceans are absorbing about half the excess carbon dioxide emitted by humanity. That effect is not strong enough to stop the rise in the atmospheric carbon dioxide concentration, but it is enough to slow it.

Figure 3-22 THE RISING GLOBAL TEMPERATURE

Temperature change (°C) compared to 1951–80 average

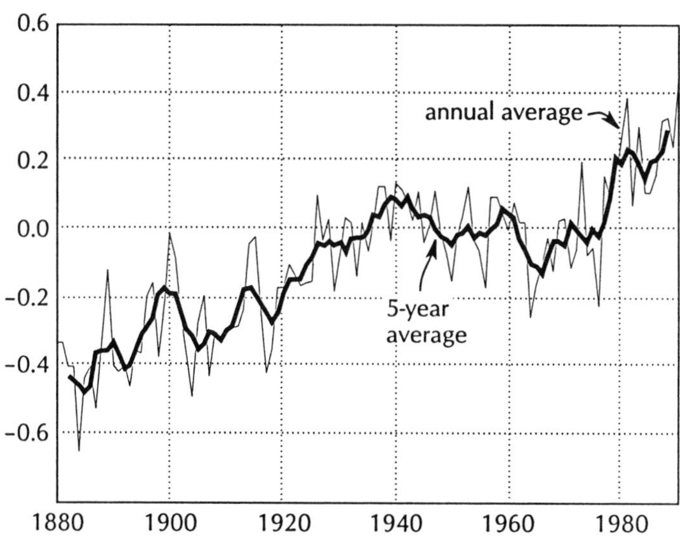

The causes and the long-term prospects for global temperature rises are still the subject of scientific and political debate. But there is little doubt that the global average temperature has been rising. The 1980s saw the six hottest years of the past century. (*Source*: *T. A. Boden et al.*)

There may also be destabilizing positive feedback loops, which, as the temperature rises, will make things even warmer. For example, as warming decreases the snow cover, the earth will reflect away less heat from the sun, thereby warming further. Melting tundra soils could release huge amounts of frozen methane, a greenhouse gas, which will cause more warming, more melting, and the release of still more methane.

No one understands how the many possible negative and positive feedback responses to the rise in greenhouse gases will interact or whether the positive or the negative feedbacks will dominate. Scientists only know that there have been temperature upheavals on earth in the past, and that they have not been quickly self-correcting or smooth or orderly. In fact they have been chaotic.

Figure 3-23 shows a 160,000 year history of Earth's temperature and of the atmospheric concentrations of two greenhouse gases, carbon dioxide and methane.[55] Temperature and greenhouse gases have varied together, though it's not clear which causes which. Most probably each causes the other in a complicated set of feedback loops.

But the most important message in Figure 3-23 is that *current* atmospheric concentrations of carbon dioxide and methane are *far higher than they have been for 160,000 years.* Whatever the consequences might be, there is no question that humanity's emissions of greenhouse gases are filling up atmospheric sinks much faster than the planet can empty them. There is a significant disequilibrium in the global atmosphere, and it is getting exponentially worse. The processes set in motion by this disequilibrium move slowly, as measured by human time scales. It may take decades for the consequences to be revealed in melting ice, rising seas, changing currents, shifting rainfall, greater storms, and migrating insects or birds or mammals. If human beings decide they do not like these consequences, it may take centuries to undo them.

The pollution emissions we have discussed in this chapter are not necessary. Pollution is not a sign of progress. It is a sign of inefficiency and carelessness. As industries realize that, they are finding ways to reduce pollution emissions by rethinking manufacturing processes from beginning to end, using "clean technology" and "precautionary pollution prevention." A circuit-board manufacturer (Aeroscientific) invests in ion-exchange columns to reclaim heavy-metal wastes and ends up with an income from the recycled metals, a much-reduced water bill, and lower liability insurance. A manufacturing company (3M) reduces its air pollution emissions, its water pollution emissions, its water requirements, and its solid waste production, and saves $200 million a year in operating expenses. An electronics company (Intel) changes its solders and fluxes to make washing with CFC solvents unnecessary, thereby reducing its contribution to ozone layer depletion and the greenhouse effect and saving $1 million per year.

Redesign for pollution prevention can be cost effective, even though the market doesn't count environmental costs. An OECD survey of over 600 French clean-technology projects found that 67% saved money on raw materials, 65% conserved water, and 8% cut energy use.

Of 45 clean technology projects in the Netherlands, 20 saved money; the others didn't affect costs one way or another.[56]

The idea of pollution prevention is not yet sweeping the industrial world, however. In Europe 80% of industrial investment in environmental protection is still in "end of the pipe" cleanup technology; only 20% is in manufacturing redesign.[57]

The potential of human ingenuity to reduce pollution flows is just beginning to be tapped. If the average lifetime of each product flowing through the human economy could be doubled, if twice as many materials could be recycled, if half as much material needed to be mobilized to make each product in the first place, that would reduce the throughput of materials by a factor of 8. If energy use became more efficient, if renewable energy sources were used, if land, wood, food, and water were used less wastefully and forests were restored, that would stop the rise of greenhouse gases and of many other pollutants. Bringing flows of energy and materials below their source limits automatically reduces pollution flows.

The stocks of some pollutants in the atmosphere, waters, and soils are falling, because human beings have taken strong action to reduce them. Some pollutants are being maintained at roughly stable levels, with great effort. Some cleanup efforts are simply moving pollutants from one sink to another, from air to land, or from water to air. Many pollutants are still rising exponentially, because the human population, its capital plant, and the materials and energy flowing through them, are still rising. Concerted attacks on the most dangerous contaminants can help steer the world away from environmental limits, provided humanity learns about them and acts in time. Greater material and energy efficiencies are even more effective than after-the-fact cleanup. They could reduce many pollution flows by factors of 2 or 5 or 10 and help bring the world back from beyond its environmental limits.

Beyond the Limits to Throughput

The evidence we have given in this chapter, the evidence contained in the world's data bases, and the evidence of the senses of just about any person looking at the world from nearly any vantage point show

Figure 3-23 GREENHOUSE GASES AND GLOBAL TEMPERATURE OVER THE PAST 160,000 YEARS

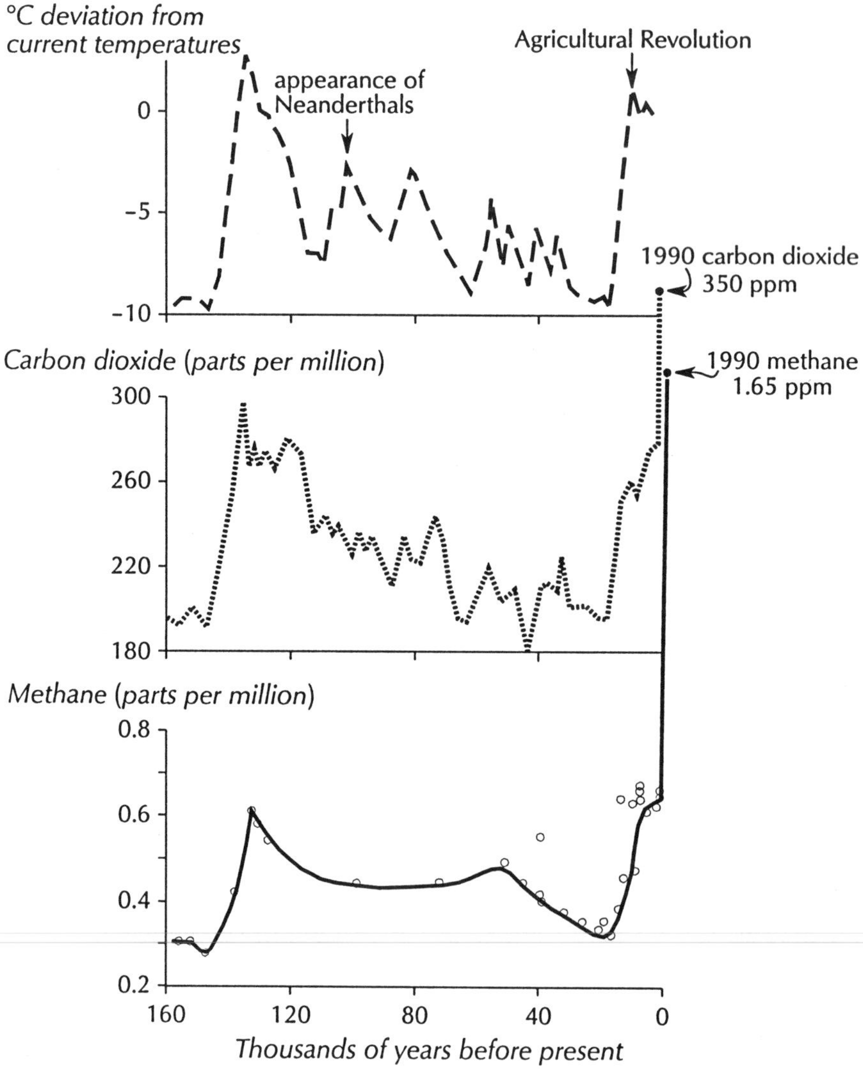

Ice core measurements show that there have been significant temperature variations on Earth (ice ages and interglacial periods) and that carbon dioxide and methane levels in the atmosphere have varied in concert with global temperature. Recent concentrations of these greenhouse gases have soared much higher than they have been since long before the appearance of the human species. (*Sources*: *R. A. Houghton et al.*)

that the human race is not using the earth's sources and sinks sustainably. Soils, forests, surface waters, groundwaters, wetlands, and the diversity of nature are being degraded. Even in places where renewable resources appear to be stable, such as the forests of North America or the soils of Europe, the quality, diversity, or health of the resource is in question. Deposits of minerals and fossil fuels are being drawn down. There is no plan and no sufficient capital investment program to power the industrial economy after the fossil fuels are gone. Pollutants are accumulating; their sinks are overflowing. The chemical composition of the entire global atmosphere is being changed.

If only one or a few resource stocks were falling while others were stable or rising, one might argue that growth could continue by the substitution of one resource for another (though there are limits to such substitution). But when many sources are eroding and many sinks are filling, there can be no doubt that human withdrawals of material and energy have grown too far. They have overshot their sustainable limits.

The limits, let us be clear, are to *throughput.* They are *speed limits,* not space limits, limits to flow rates, not limits to the number of people or the amount of capital (at least not directly). To be beyond them does not mean running into an absolute wall. It may even mean that material and energy throughputs can still grow for a while, before negative feedbacks from overstressed sources or sinks force them down. But down is the direction that throughputs will have to go, by human choice or by strong and unpleasant natural feedbacks.

Many people recognize at least on a local level that throughputs to the human economy have grown beyond limits. Los Angeles emits more air pollution than human lungs can bear. The forests of the Philippines are nearly gone. The soils of Haiti have been worn down in places to bare rock. The chemical load in the Rhine is so high that dredged silt from Dutch harbors has to be treated as hazardous waste. In the case of particular problems, such as the CFCs that erode the ozone layer, there has been not only recognition of an overshoot, but determined international effort to take corrective action.

But there is little discussion about the *general problem* of overshoot, little pressure for the technical changes that are urgently necessary, and almost no willingness to deal with the underlying driving forces of population and capital growth. Even concerned, informed groups like the World Commission on Environment and Development, which looked hard at world trends and labelled them "simply unsustainable," have not been able to say, straightforwardly, "the human world is beyond its limits," or to grapple seriously with the question of reducing throughput.

The reasons for general avoidance of the issue of limits are political. Any talk of limits feeds into a bitter argument, already underway, about who are the real polluters of the world. The throughput per rich person is much greater than the throughput per poor person. The throughput per ton of steel is greater for an inefficient plant in East Europe than for a superefficient minimill in Japan. One Swiss, it is said, uses as much energy as forty Somalis, and one Russian uses as much energy as one Swiss without even getting a decent standard of living out of it. If the world as a whole is exceeding its limits, who should do something about it: the wasteful rich or the multiplying poor or the sloppy ex-socialists?

As far as the planet is concerned, the answer is all of the above. Environmentalists sometimes summarize the causes of environmental deterioration with a formula they call IPAT:

Impact = Population 3 Affluence 3 Technology

The impact (throughput) of any population or nation upon the planet's sources and sinks is the product of its population (P) times its level of affluence (A) times the damage done by the particular technologies (T) that support that affluence.[58]

Since each term in this equation multiplies impact equally, it would follow that every society should make improvements where it has the most opportunity to do so. The South has the most room for improvement in P, the West in A, the East in T.

The total scope for improvement is astonishing. If we define each term in the IPAT equation more precisely, we can see how many ways there are to reduce throughput, and what great reductions are possible (see Table 3-4).[59]

Affluence can be defined as capital stock per person—the number of cups (or television sets or cars or rooms in a house) per person. The impact or *throughput* due to affluence consists of the material flows needed to maintain each form of capital. For instance, if there are three china cups per person, maintaining those cups takes water and soap to wash them and a small flow of cups to replace annual breakage. If a person uses and discards polystyrene hot cups for coffee at work, the maintenance flow includes all the cups used in a year and the petroleum and chemicals needed to make the polystyrene.

The impact of technology is defined in Table 3-4 as the energy needed to make and deliver each material flow, multiplied by the environmental impact per unit of energy. It takes energy to mine the clay for a ceramic cup, and to fire the clay, and to deliver the cup to the household, and to heat the water to wash it. It takes energy to find and pump the oil for polystyrene cups, and to transport the oil, run the refinery, form the polymer, mold the cups, deliver the cups, and transport the used cups to the dump. Each kind of energy has its environmental impact. The impact can be changed technologically with pollution control devices, with energy efficiency changes, or by switching to another energy source.

Changes in any factor in Table 3-4 will bring the human economy closer to or further from the earth's limits. Reducing population or the stock of material accumulated by each person will help keep the human world within the limits of the planet. So will lower rates of energy or material flows for replacement and maintenance. So will less pollution impact per unit of material or energy. Table 3-4 lists some of the tools that might help reduce each factor in the equation and also some guesses about how much each factor might be reduced, and over what time scale.

You can see, when the options are laid out in this way, that there are many, many choices. Human impact on the planet's sources and sinks could be reduced to an astounding degree. Even if you believe only the lower bound for each estimate of possible change, taken all together they could reduce the human impact on the planet by a factor of a *thousand or more*—possibly much more.

As MIT economist Lester Thurow has said, "If the world's population had the productivity of the Swiss, the consumption habits of the

Table 3-4 THE ENVIRONMENTAL IMPACT OF POPULATION, AFFLUENCE, AND TECHNOLOGY

Population	*Affluence*		*Technology*	
Population	3 Capital stock / Person	3 Material throughput / Capital stock	3 Energy / Material throughput	3 Environmental impact / Energy
Example				
Population	3 Cups / Person	3 Water + soap / Cups/year	3 Gigajoules or kilowatt-hours / Kilogram water + soap	3 CO_2, NOx, land use / Gigajoules or kilowatt-hours
Applicable tools				
Family planning Female literacy Social welfare Role of women Land tenure	Values Prices Full costing What do we want? What is enough?	Product longevity Material choice Minimum-materials design Recycle, reuse Scrap recovery	End-use efficiency Conversion efficiency Distribution efficiency System integration Process redesign	Benign sources Scale Siting Technical mitigation Offsets
Approximate scope for long-term change				
~2 3	?	~3–10 3	~5–10 3	~10^2–10^{3+} 3
Time scale of major change				
~50–100 years	~0–50 years	~0–20 years	~0–30 years	~0–50 years

Chinese, the egalitarian instincts of the Swedes, and the social discipline of the Japanese, then the planet could support many times its current population without privation for anyone. On the other hand, if the world's population had the productivity of Chad, the consumption habits of the United States, the inegalitarian instincts of India, and the social discipline of Argentina, then the planet could not support anywhere near its current numbers."[60]

If there are so many options, why is the present world not going to much trouble to pursue any of them? What if it did? What would happen if population, affluence, and technology trends begin to turn around? What about the ways they are interconnected to each other? What happens if throughput is reduced by technical change, but then population and capital grow still further? What happens if throughput isn't reduced at all?

These are questions not about resources viewed separately, as we have seen them in this chapter, but about resources viewed together, interacting with population and capital, which are in turn interacting with each other. To address them we need to move on from a static, one-factor-at-a-time analysis to a dynamic whole-system analysis.

chapter 4

The Dynamics of Growth in a Finite World

Man is like every other species in being able to reproduce beyond the carrying capacity of any finite habitat. Man is like no other species in that he is capable of thinking about this fact and discovering its consequences.

William R. Catton, Jr.[1]

In most parts of the world capital is growing faster than population; in a few parts of the world the reverse is true. In some places increasing economic security and empowerment are bringing birth rates down, in other places poverty and social disintegration are bringing death rates up. People who are getting richer are demanding more industrial products, more energy, cleaner air. Poor people are struggling for clean water, land to farm, firewood to burn. Some technologies are increasing flows of pollution, others are decreasing them. Nonrenewable resource stocks and some renewable ones are being depleted; others are being utilized more extensively and efficiently.

Powerful trends are running counter to each other. To put them together and glimpse their combined implications, we need a model more complex than the ones in our heads. This chapter is about the computer model we have used, its structure, and its basic findings.

The Purpose and Structure of World3

First a word about models. A model is any simplified representation of reality. Every word in this book is a model. "Growth," "population," "forest," "water," are just symbols, which stand for very complex realities. Every graph, chart, map, and piece of data about the world is a model. To say the world's population has reached 5.4 billion is not exactly accurate, nor does it begin to capture the actuality, the diversity, or humanity of those people. The way we have put words and numbers together to make this book is a model of what is in our minds. It is our best attempt to symbolize our thoughts, but it is only a model of those thoughts. And of course our thoughts, and every person's thoughts, are only models of the real world.

Therefore we have a difficulty. We are about to talk about a formal model, a computer-based simulation of the world. For this model to be of any use, we will have to compare it to the "real world," but neither we nor our readers have an agreed-upon real world to compare it to. We only have the worlds of our mental models. Mental models are informed by objective evidence and subjective experience. They have allowed *Homo sapiens* to be a tremendously successful biological species. They have also gotten people into all kinds of trouble. But whatever their strengths and weaknesses, human mental models must be ludicrously simple compared with the immense, complex, ever-changing universe within which they exist.

To remind ourselves of our inevitable dependence upon models, from here on we will put the World3 model's referent, the "real world," in quotation marks. What we mean by "real world" or "reality" is the mental models of the authors of this book. We make no apology for that referent; it's the only one we have. Mental models are the only referents that any author or reader has. We can't escape that fact, and it's important to acknowledge it.

World3 is not a difficult model to understand. It keeps track of stocks such as "population" and "industrial capital" and "pollution" and "cultivated land." Those stocks change through flows such as "births" and "deaths" (in the case of population), "investment" and "depreciation" (in the case of capital), and "pollution generation" and "pollution assimilation" (in the case of pollution). Land cultivated multiplied by average land yield gives total food production. Food production divided by population gives food per capita. The amount of food per capita influences the death rate.

Nothing remarkable here. The components of World3 are quite ordinary. But they are put together in a way that is dynamically complex. World3 takes into account the momentum of population growth, the accumulation of pollution, the long lifetime of capital plant, the changing flows of resources, the competing pulls for investment. It focuses especially on the time it takes for things to happen and on the delays in flows of information and physical processes. It is based on feedback processes, which means that an element can be the partial cause of its own future behavior. A change in population, for example, may cause a change in the economy, which may then cause another change in population.

Many causal relationships in World3 are *nonlinear*—they are not straight lines, not strictly proportional over all ranges of the related variables. For example, in World3 more food per capita causes an increase in human lifetime, but not a linear one. Figure 4-1 shows a plot of food per capita versus life expectancy. If people who are inadequately nourished get more food, their life expectancy can increase greatly. But more food for a population that is already well fed has little effect on life expectancy (and at some point may actually decrease it).

Nonlinear relationships are found throughout the "real world" and throughout World3. Because of them both the "real world" and World3 can sometimes produce surprising behavior, as we'll demonstrate later in this chapter.

World3's nonlinearities and feedback structure make it dynamically *complex*, but the model is not *complicated.* It does not distinguish among different geographic parts of the world, nor does it represent separately the rich and the poor. It keeps track of only one generic pollutant, which moves through and affects the environment in ways that are typical of the hundreds of pollutants science has identified. It distin-

Figure 4-1 NUTRITION AND LIFE EXPECTANCY

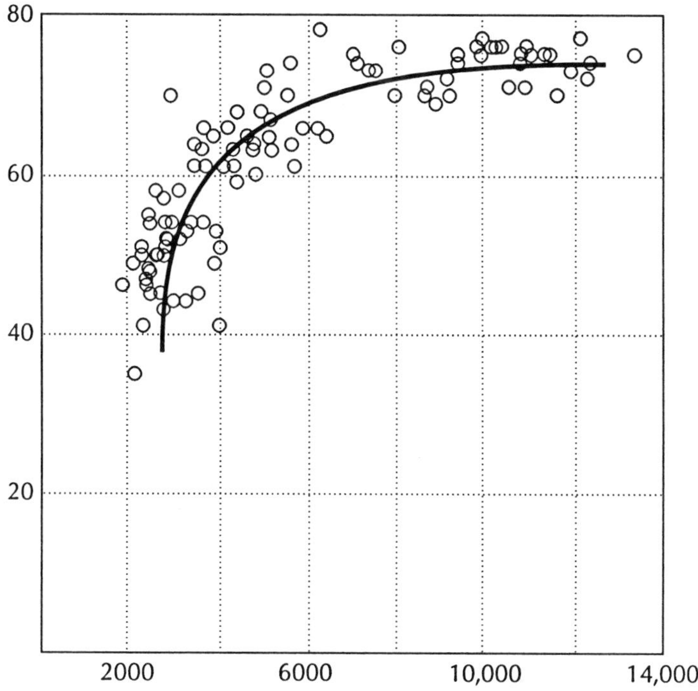

The life expectancy of a population is a nonlinear function of the nutrition that population receives. Each point on this graph represents the average life expectancy and nutritional level of one nation in 1988. Nutritional level is expressed in vegetable calorie equivalents per person per day: calories obtained from animal sources are multiplied by a conversion factor of 7 (since about 7 calories of vegetable feed are required to produce 1 calorie of animal origin). *(Sources: Food and Agriculture Organization; Population Reference Bureau.)*

guishes the renewable resources that produce food from the nonrenewable ones that produce fossil fuels and minerals, but it doesn't keep separate account of each type of food, each fuel, each mineral.

That degree of simplicity surprises some people who assume that a world model ought to contain everything one knows about the world,

Figure 4-2 POSSIBLE MODES OF APPROACH OF A POPULATION TO ITS CARRYING CAPACITY

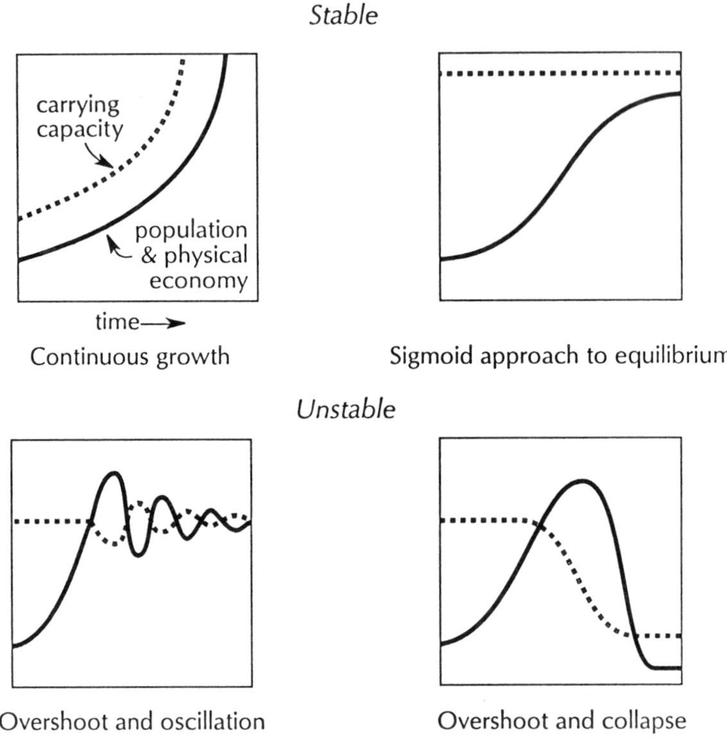

The central question addressed by the World3 model is: Which of these behavior modes is likely to be the result as the human population and economy approach their carrying capacity?

especially all the distinctions that are so fascinating and, from the point of view of each individual, so critical. In fact even large, highly detailed computer models must be gross simplifications of human knowledge. Computer modelers, if they are not to produce impenetrable thickets, have to discipline themselves. They cannot put into their models all they know; they have to put in only what is relevant for their purpose. The art of modeling, like the arts of poetry writing or architecture or engineering design, is to include just what is necessary to achieve the purpose, and no more. That is easy to say and hard to do.

Therefore to understand a model and judge its "validity," one needs to understand its purpose. The purpose—the *only* purpose—of World3 is to understand the possible modes of approach of the human economy to the carrying capacity[2] of the planet. It is for that purpose *only* that we believe the model is "valid." There are many other important questions to ask—what are the development possibilities for Africa, how to design a family planning program, how to close the gap between the rich and the poor. Computer models, different ones from World3, can help answer some of those questions. All of them should be answered, we believe, within the context of a global society encountering and adapting to the earth's limits.

A growing population can approach its carrying capacity in four generic ways (see Figure 4-2). It can grow without interruption, as long as its limits are far away or growing faster than it is. It can level off smoothly into a balance with the carrying capacity, in a behavior that ecologists call logistic or S-shaped or *sigmoid* growth. It can overshoot its limits and oscillate around them for a while before leveling off. Or it can overshoot its limits, destroy its resource base, and crash.

We created World3 to answer two questions: *Which of the behavior modes shown in Figure 4-2 is most likely to be followed by the human population and economy? What conditions or policies will increase the chances of a smooth approach to planetary limits?*

Those are questions about broad behavioral possibilities, not precise future conditions. Answering them requires a different kind of modeling and a different kind of information than does precise prediction. For example, if you throw a ball straight up into the air, you know enough to forecast what its general behavior will be. It will rise with decreasing velocity, then reverse direction and fall with increasing velocity until it hits the ground. You know it will not continue rising forever, nor begin to orbit the earth, nor loop three times before landing.

If you wanted to predict exactly how high the ball will rise or precisely where and when it will hit the ground, you would need precise information about the ball, the altitude, the wind, the force of the initial throw. Similarly, if we wanted to predict the exact size of the world population in 2026, or when world oil production will peak, or which limit will affect a specific nation first, we would need a very complicated model—in fact an impossible one.

It is not possible to make accurate "point predictions" about the future of the world's population, capital, and environment. No one knows enough to do that. And the future of that system is too dependent on human choice to be precisely predictable. It is possible, however, and critically important, to understand the broad behavioral possibilities of the system, especially since collapse is one of them. Therefore we put into World3 the kinds of information one uses to understand the generic behavior modes of thrown balls, not the kinds of information one would need to describe the exact trajectory of one particular throw of one specific ball.

So, for example, we thought it important to represent pollution as something that is generated by agriculture and industry and something that can affect the health of human beings and crops. We included a delay before pollution finds its way to a place where it can do measurable harm, because we know that it takes time for a pesticide to work its way down into groundwater, or for a chlorofluorocarbon molecule to rise up and damage the ozone layer, or for mercury to wash into a river and accumulate in the flesh of fish. We wanted to represent the fact that natural processes can render most pollutants harmless after awhile, and also the fact that those natural pollution cleanup processes can themselves be impaired. All those general characteristics of pollution are in World3, but the model does not distinguish PCBs from CFCs from DDT.

We included in World3 the best numbers we could find, but we acknowledge a large range of uncertainty around some numbers. Because of all the uncertainties and simplifications we know are in the model (and others we don't know are in the model) we do not put faith in the exact numerical path the model churns out for population or pollution or capital or food production. We do trust the basic behavior, the fact that the population or pollution grows, or holds steady, or oscillates, or declines. We think the primary interconnections in World3 are "valid," and those interconnections determine the model's general behavior.

What are those primary interconnections? They begin with the feedback loops around population and capital, which we described in Chapter 2, and which are reproduced in Figure 4-3. They give population and capital the potential to grow exponentially if the positive birth

Figure 4-3 FEEDBACK LOOPS GOVERNING POPULATION AND CAPITAL GROWTH

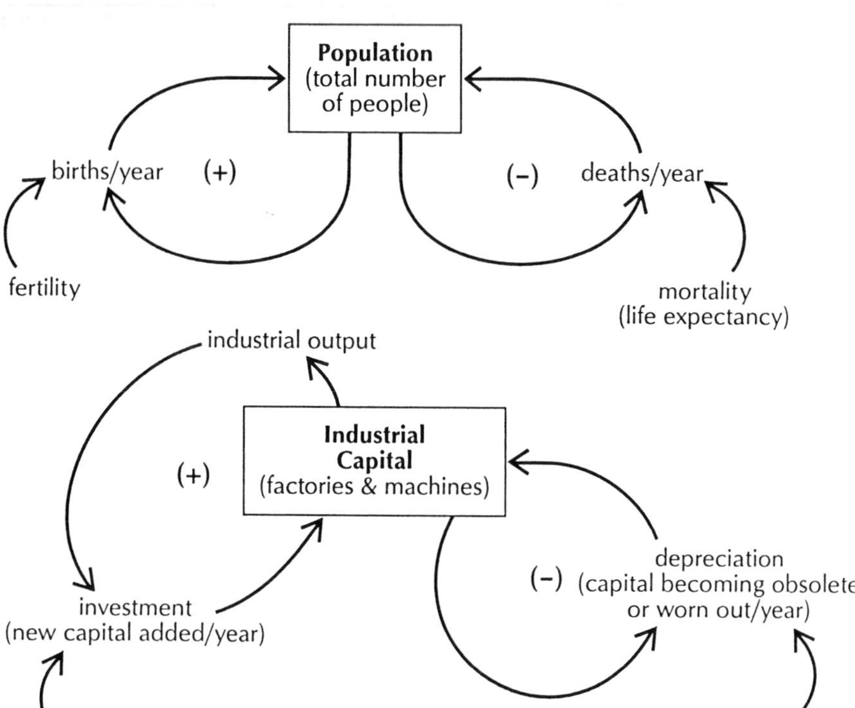

The central feedback loops of the World3 model govern the growth of population and of industrial capital. The two positive feedback loops involving births and investment generate the exponential growth behavior of population and capital. The two negative feedback loops involving deaths and depreciation tend to regulate this exponential growth. The relative strengths of the various loops depend on many other factors in the system.

and investment loops dominate, the potential to decline if the negative death and depreciation loops dominate, and the potential to stay in equilibrium if the loops are balanced.

In all our feedback-loop diagrams, such as Figure 4-3, the arrows indicate simply that one variable influences another through physical or informational flows. The *nature* and *degree* of influence are not shown on the diagrams, though of course they have been specified mathematically for the computer.

The boxes in the diagrams indicate stocks. These are accumulations of physical quantities, such as population or factories or pollution. The stocks in a system tend to change only slowly because the things they contain have long lifetimes. The stocks represent the present state of the system as accumulated during its past history. The factories in place, the number of people, the concentration of pollutants, the remaining resources, the developed land, all are important stocks in the World3 system. They determine the limitations and the possibilities of the system at each moment of simulated time.

Feedback loops in the causal diagrams are marked with (+) if they are positive loops—self-reinforcing loops that can generate exponential growth. They are marked with (−) if they are negative loops—goal-seeking loops that reverse the direction of change or try to pull the system into balance or equilibrium.

Some of the ways population and capital influence each other in World3 are shown in Figure 4-4. Industrial capital turns out many products, one of which is agricultural inputs—tractors, irrigation pumps, fertilizers, pesticides. Agricultural inputs and cultivated land determine food production. Food is also affected by pollution, which comes from both industrial and agricultural activity. Food per person influences the mortality of the population.

Figure 4-5 shows some of the links in World3 connecting population, capital, services, and nonrenewable resources. Some industrial output takes the form of service capital—houses, schools, hospitals, banks, and the equipment they contain. Output from service capital divided by the population gives the average level of services per person. Health services decrease the mortality of the population. Education and family planning services bring down the birth rate. Rising industrial output per person is also assumed to decrease fertility directly (after a delay) by changing employment patterns, the costs of raising children, the benefits of raising children, and the ways families allocate their time.

Each unit of industrial output is assumed to consume nonrenew-

Figure 4-4 FEEDBACK LOOPS OF POPULATION, CAPITAL, AGRICULTURE, AND POLLUTION

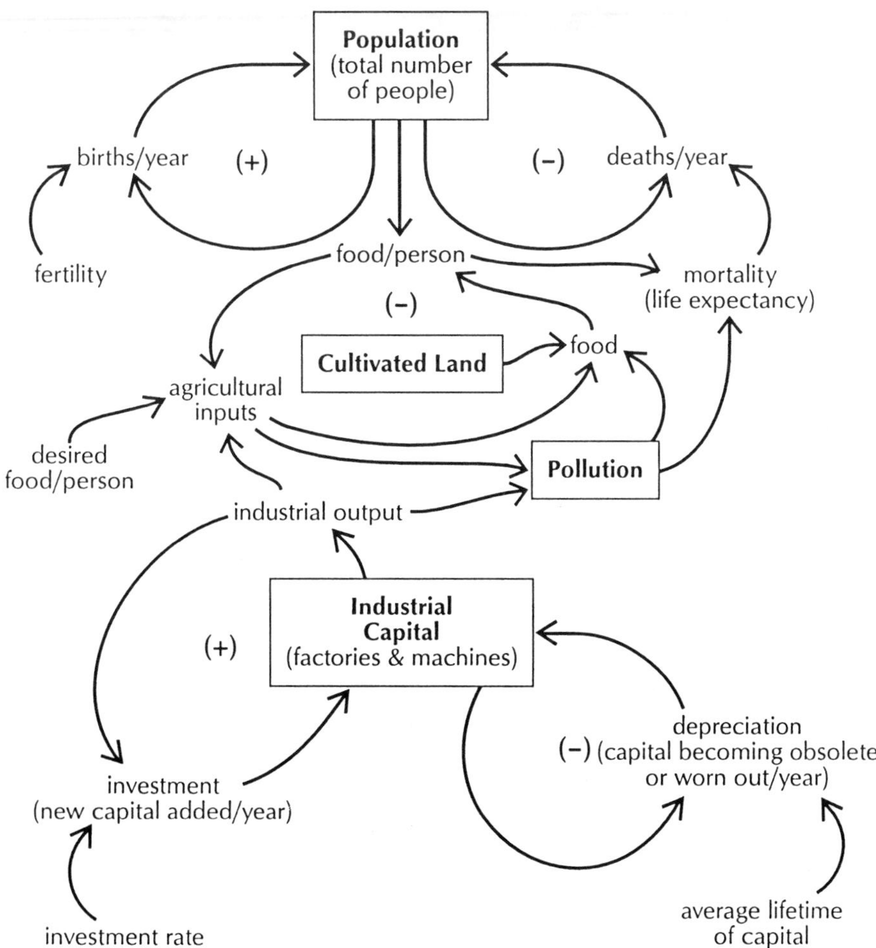

Some of the interconnections between population and industrial capital operate through agricultural capital, cultivated land, and pollution. Each arrow indicates a causal relationship, which may be immediate or delayed, large or small, positive or negative, depending on the assumptions included in each model run.

Figure 4-5 FEEDBACK LOOPS OF POPULATION, CAPITAL, SERVICES, AND RESOURCES

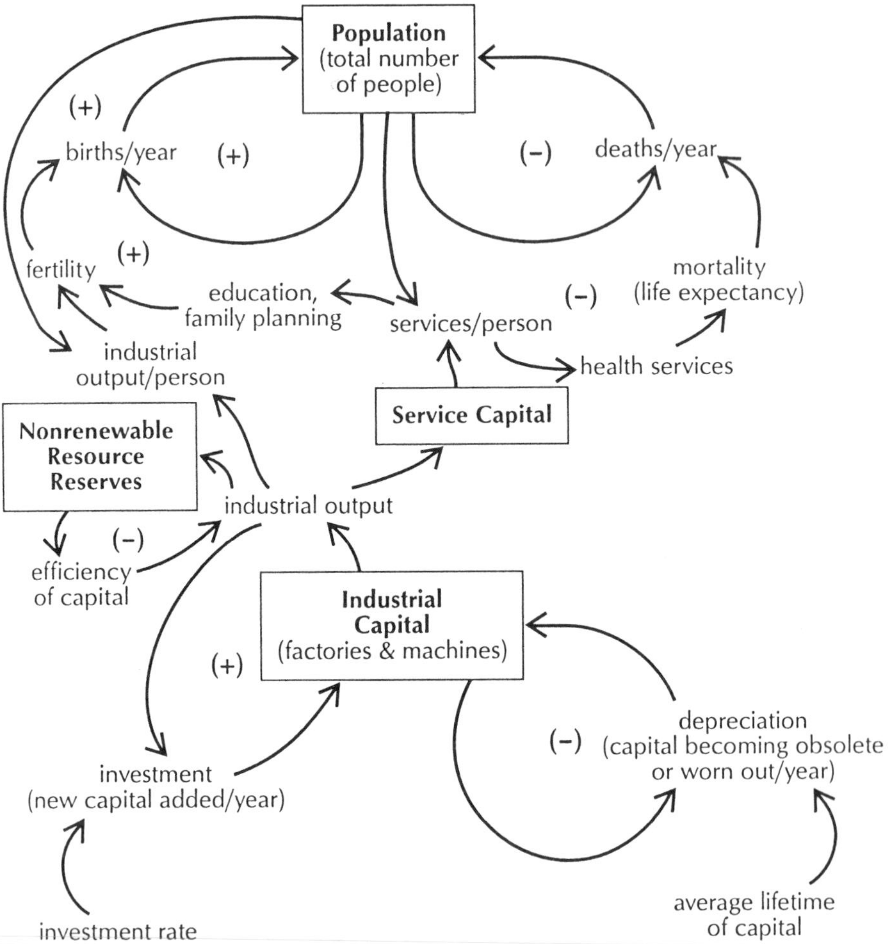

Population and industrial capital are also influenced by the levels of service capital (such as health and education services) and of nonrenewable resource reserves.

able resources. Exactly how many resources are needed per unit of industrial production can be changed by technology, but the model does not allow industry to make something out of nothing. As nonrenewable reserves diminish, the grade of the remaining reserves is assumed to decline and deposits are assumed to get deeper and farther away from their places of use. That means that as depletion proceeds, more capital is necessary to extract a ton of copper or a barrel of oil from the earth.

The relationship between resources remaining and capital required to obtain them is highly nonlinear. Some clues about how it looks are shown in Figure 4-6, which shows the energy necessary to extract iron and aluminum at various ore grades. Energy is not capital (actual capital is hard to measure), but energy implies capital, since as the grade goes down, more rock must be lifted per ton of final resource, the rock must be crushed into finer particles, it must be sorted more accurately into its component minerals, and larger tailings piles must be heaped up, all of which is done by machines. If more capital is needed in the resource-producing sector, less investment is available for other purposes in the economy.

Diagrams of all the interconnections in World3 are shown in the Appendix. It is not necessary to understand every one of these interconnections to comprehend how the model works. The most important features of the model are:

- the growth processes
- the limits
- the delays
- the erosion processes

We have already described the growth processes of population and capital in Chapter 2. Limits in the "real world" we discussed in Chapter 3. Limits as they are represented in World3, delays, and erosion we will describe next. The important question you should keep in mind throughout the following discussion is whether and under what conditions there are parallels between the computer model we are discussing and the "real" population and economy, as far as you know them through your own mental model.

Limits and No Limits

An exponentially growing economy taking resources from and emitting wastes into a finite environment begins to stress that environ-

Figure 4-6 ENERGY REQUIRED TO PRODUCE PURE METAL FROM ORE

Thousand kilowatt-hours per ton of metal

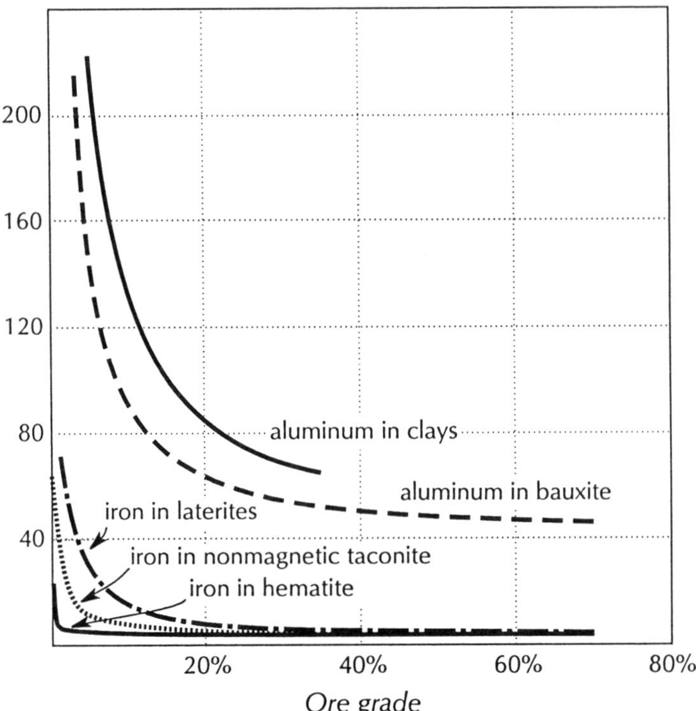

As their metal content declines, ores require increasingly large amounts of energy for their purification. (*Source: N. J. Page and S. C. Creasey.*)

ment long before reaching ultimate limits. The environment then begins to send signals and pressures to the growing economy—signals of resource scarcity, pressures from accumulating wastes. Those signals and pressures are negative feedback loops. They seek to bring the economy into alignment with the constraints of the surrounding system. That is, they seek to stop its growth.

World3 contains just four kinds of physical and biological limits. All of them can be raised or lowered by actions, changes, and choices within the model world. These limits are:

- *Cultivable land,* which can be increased up to a limit of 3.2 billion hectares by investment in land development. The cost of developing new land is assumed to rise as the most accessible and favorable land is developed first. Land can also be removed from production by erosion and urbanization.
- *The yield achievable on each unit of land,* which can be raised by inputs like fertilizer. These inputs have diminishing returns; each additional kilogram of fertilizer produces less additional yield than the kilogram before. We assume that the upper yield limit is a worldwide average of 6500 kilograms of grain per hectare, equivalent to the highest yields obtained by single countries today. World3 also assumes that land yield can be reduced by pollution.
- *Nonrenewable resources* like minerals and fossil fuels. We assume there are enough of these resources to supply 200 years worth of extraction at 1990 extraction rates. The capital cost of finding and developing nonrenewable resources is assumed to rise, as the richest and most convenient deposits are exploited first.
- *The ability of the earth to absorb pollution,* which is assumed to erode as pollution accumulates, and which can regenerate itself if the pollution load decreases. Quantitatively this is the least-known limit of all. We assume that if pollution rises to 10 times its 1990 global level, it would reduce human lifetime by only 3% and accelerate the degradation of land fertility by 30% (and then we test other estimates in the model to see what their effects would be).

Of course in the "real world" there are other kinds of limits, including managerial and social ones. Some of them are implicit in the numbers in World3, since those numbers came from the world's "actual" history over the past ninety years. But World3 has no explicit war, no labor strikes, no corruption, no trade barriers, and its simulated population does its best to solve perceived problems, undistracted by struggles over political power. Since it lacks many social limits, World3 may well paint an overly optimistic picture of future options.

What if we're wrong about, say, the amount of resources still to be discovered? What if the actual number is only half of what we've assumed, or double, or 10 times more? What if the earth's "real" ability to absorb pollution without harm to the human population is not 10 times the 1990 rate of emission, but 50 times or 500 times? (Or 0.5 times?)

A computer model is a device for making tests, and all those "what ifs" are testable. It is possible, for example, to set the numbers on World3's limits astronomically high. It is even possible to program them to grow exponentially. We have tried that, and so have others.[3] When all limits are removed from the model system by an assumed technology that is unlimited and cost- and error-free, the simulated human economy grows for as long as we let the model run. Figure 4-7 shows what happens. Population growth slows and begins to level off in this model run (at about 15 billion) because of the demographic transition. Industrial output grows right off the top of the graph. In the simulated year 2100 the global economy is producing 55 times as much industrial output and 8 times as much food as it did in 1990. This growth is achieved while the world uses only 5% as many resources and generates only 15% as much pollution as it did in 1990. To achieve this unrealistic outcome the world would have to accumulate more than 60 times as much productive capital in the twenty-first century as it did in the twentieth century.

This run tells you something about World3, something about modeling, and very little, we believe, about the future of the "real world." What it reveals about World3 is that the model has built into its structure a self-limiting constraint on population. Population will eventually level off, if industrial output per capita rises high enough. But the model contains no self-limiting constraint on capital. We see little "real-world" evidence that the richest people or nations have lost interest in getting richer. Therefore we have assumed that capital-owners will continue to try to multiply their wealth indefinitely and that consumers will continue to be willing to increase their consumption. That assumption can and will be changed in policy runs presented in Chapter 7.

Figure 4-7 also demonstrates one of the most basic principles of modeling: Garbage In, Garbage Out or GIGO. The computer will tell you the logical consequences of your assumptions, but it will not tell you whether your assumptions are logical. If you assume the earth is

How to Read World3 Scenarios

In Chapters 4, 6, and 7 of this book we will show 14 different "computer runs" or scenarios generated with World3. Each run starts with the same basic model structure and changes some numbers to test different estimates of "real world" parameters, or to incorporate more optimistic projections of the development of technologies, or to see what happens if the world chooses new policies, ethics, or goals.

When the new numbers are entered, we use World3 to calculate the interactions among all its 225 variables. The computer calculates a new value for each variable every six months in simulated time from the year 1900 to the year 2100. The model thus produces more than 90,000 numbers for every scenario. We can't possibly picture all this information; we have to consolidate and simplify it to understand it ourselves and to convey it to you.

We do that by plotting out on time graphs the values of only a few key variables, such as population, pollution, and natural resources. For this book we provide two such graphs for each scenario. The format is the same for each scenario we show.

The top graph, called "State of the World," is indicative of the total burden on the planet. It will show global totals for:

population
food production
industrial production
relative level of pollution (1970 equals 1)
remaining nonrenewable resources

The second graph, called "Material Standard of Living," is indicative of average individual human welfare. It will show values for:

food production per person
consumption goods per person
average life expectancy
service output per person

We have omitted the numerical values of the vertical scales for these 9 variables, since their precise values at each point in time are not meaningful. Instead you should notice how the shapes of the curves change from one set of assumptions to another. To facilitate that comparison, we have kept the vertical scales for each of the 9 variables identical across Figure 4-7 and all 13 scenarios (most of which appear in Chapters 6 and 7). However, 2 variables on the same graph may be plotted on very different scales with different units. For example, the scale for population goes from 0 to 13 billion people, while the scale for life expectancy goes from 0 to 90 years.

infinite and that human desires are unquenchable, World3 will give you infinite growth. We label the run shown in Figure 4-7: Infinity In, Infinity Out, abbreviated IFI-IFO (pronounced "iffy-iffo"). The important question about this and every other computer run is not only whether you believe the model behavior, but also whether you believe the driving assumptions that produce that behavior, in this case the assumption of a boundless earth.

We don't believe the run shown in Figure 4-7. Under what we would call more "realistic" assumptions, the model begins to show the behavior of a growing system running into resistance from physical limits.

Limits and Delays

A growing physical entity will stop exactly at its limits (sigmoid or S-shaped growth) only if it receives accurate, prompt signals telling it where it is with respect to its limits, and only if it responds to those signals quickly and accurately.

For example, imagine that you are driving a car and you see a stoplight turn red up ahead. You can pull the car up to a smooth halt right at the light because you have a fast, accurate visual signal telling you where the light is, because your brain responds rapidly to that signal, because your foot moves quickly as you decide to step on the brake, and because the car responds to the brake with a speed you know and have accounted for.

If your side of the windshield were fogged up and you had to depend on a passenger to tell you where the stoplight was, the short delay in communication could cause you to shoot past the light (unless you slowed down to accommodate the delay). If the passenger lied, or if

Figure 4-7 INFINITY IN, INFINITY OUT

If all physical limits to the World3 system are removed, population grows to 15 billion and levels off in a demographic transition. The economy grows until by the year 2100 it is producing 55 times the 1990 level of industrial output while using only 5% as many nonrenewable resources and producing only 15% as much pollution.

FIGURE 4-7

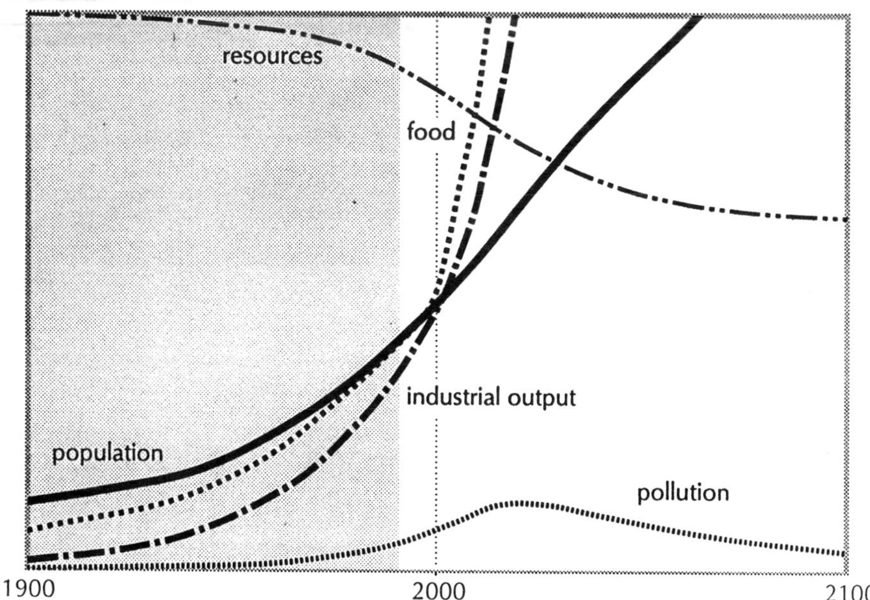
State of the world
resources
food
industrial output
population
pollution
1900
2000
2100

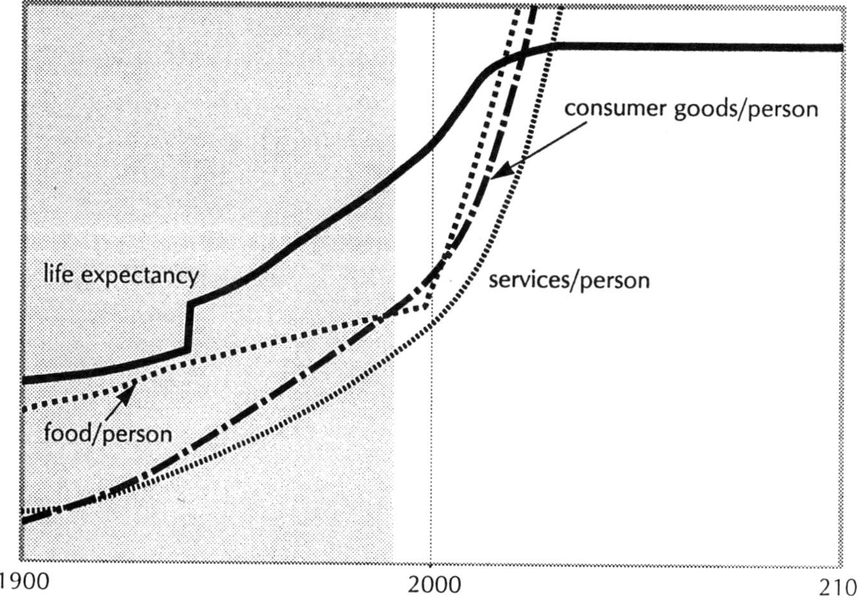
Material standard of living
consumer goods/person
life expectancy
services/person
food/person
1900
2000
2100

you denied what you heard, or if it took the brakes two minutes to have an effect, or if the car had such momentum that it took several hundred yards or meters to stop, you would be in trouble.

A system cannot come to an accurate adjustment to a limit if the controlling signal is delayed or distorted, if it is ignored or denied, or if it can respond only after a delay. If any of those conditions pertain, the growing entity will correct itself too late and overshoot (Figure 4-8).

We have already indicated where some of the information and response delays occur in World3. One of them is the delay between the time when a pollutant is released into the biosphere and the time at which it does measurable harm to human health or the human food supply. An example is the fifteen- to twenty-year lag before a chlorofluorocarbon molecule released on the earth's surface begins to degrade the stratospheric ozone layer, as described in the next chapter. Another example is the slow percolation of PCBs through the environment.

Since 1929 industry has produced some 2 million tons of the stable, oily, nonflammable chemicals called polychlorinated biphenyls, or PCBs. They have been used primarily to dissipate heat in electrical capacitors and transformers; they are found throughout the world wherever there are electrical lines, electrical equipment, and hydraulic equipment. For forty years users of these chemicals dumped them in landfills, along roads, into sewers and water bodies, without thinking of the environmental consequences. Then a landmark study in 1966, designed to detect DDT in the environment, reported that in addition to DDT, it had found widespread PCBs as well.[4] Since then PCBs have been found just about everywhere.

> PCBs are in almost every component of the global ecosystem. The hydrosphere is a major source of atmospheric PCBs. . . . PCB residues have also been detected in river, lake, and ocean sediments. . . . A comprehensive study of the Great Lakes ecosystem clearly illustrates the preferential bioconcentrations of PCB residues in the food chain.[5]

> DDT and PCBs are the only organochlorines that have been monitored on a systematic basis in arctic marine mammals. . . . The PCB levels in the breast milk of the Inuit women are among the highest ever reported. . . . A high consumption of fishes and sea mammals is probably the main route of intake for PCBs. . . . These results suggest

Figure 4-8 Structural Causes of the Four Possible Behavior Modes of the World Model

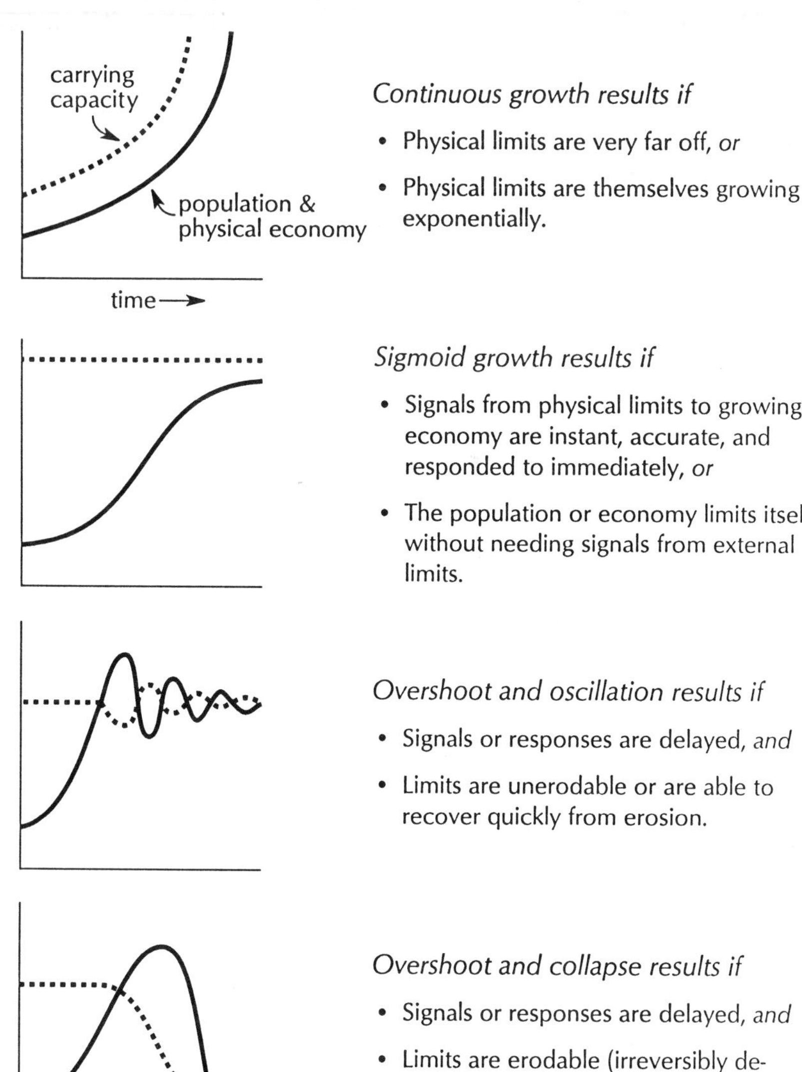

> that toxic compounds such as PCBs could play a role in the impairment of immunity and in the high occurrence of infection among Inuit children.[6]

> [In the Waddenzee on the coast of the Netherlands] the reproductive success of the seals receiving the diet with the highest level of [PCBs] was significantly decreased. . . [which shows that] the reproductive failure in common seals is related to feeding on fish from that polluted area. . . . These findings corroborate the results from experiments with mink, where PCBs impaired reproduction.[7]

Most PCBs are relatively insoluble in water but soluble in fats, and they have very long lifetimes in the environment. They move quickly through the atmosphere, and slowly through soils or sediments in streams and lakes, until they are taken up into some form of life, where they accumulate in fatty tissue and increase in concentration as they move up the food chain. They are found in the greatest amounts in carnivorous fish, sea birds and mammals, human fat and human breast milk. They interfere with immune and endocrine function, especially with reproduction and the development of the fetus.

Because of these slow-moving, long-lasting, bioaccumulating characteristics, PCBs have been called a "biological time bomb." Although PCB manufacture and use has been banned in many countries since the 1970s, a huge stock still exists. Of the total amount of PCBs ever produced about 70% are still in use or stored in abandoned electrical equipment. In countries with hazardous waste laws, there is hope that those PCBs will be disposed of by controlled incineration. The remaining 30% has been released into the environment. Only 1% has reached the oceans; that amount is causing the effects already measurable in fish, seals, birds, and people. The 29% unaccounted for is dispersed in soils, rivers, and lakes, where it may go on moving into living creatures for decades.[8]

Figure 4-9 shows another example of a pollution delay, the slow transport of chemicals through soil and into groundwater. From the 1960s until 1990, when it was finally banned, the soil disinfectant 1,2-dichloropropene (DCPe) was applied heavily in the Netherlands in the cultivation of potatoes and flower bulbs. It contains a contaminant, 1,2-

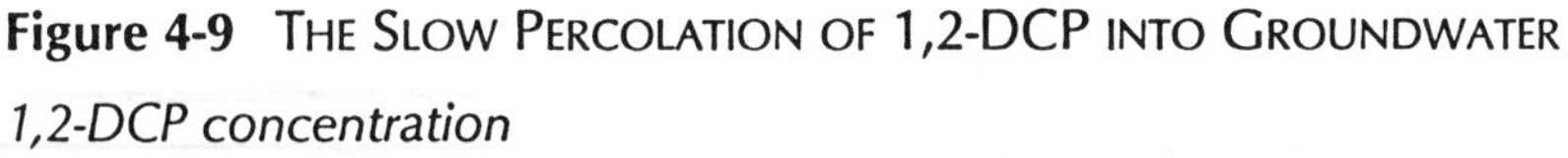

Figure 4-9 THE SLOW PERCOLATION OF 1,2-DCP INTO GROUNDWATER

1,2-DCP concentration

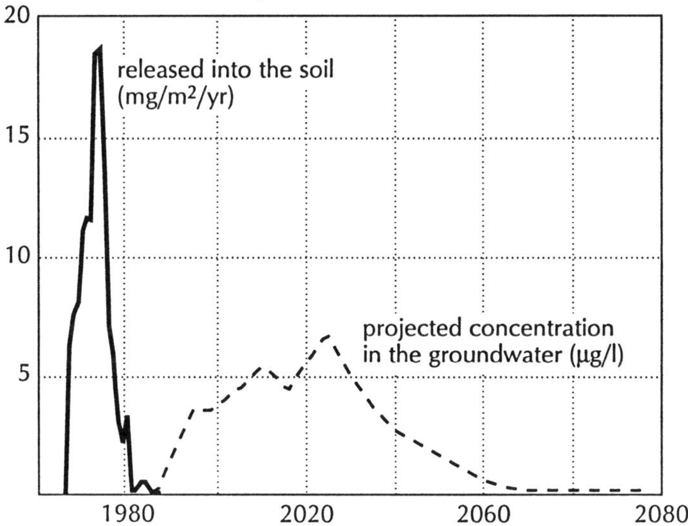

The soil disinfectant DCP was used heavily in the Netherlands in the 1970s, then it was restricted, and finally in 1990 it was banned. As a result, the concentration of DCP in the upper levels of agricultural soils has declined quickly. However, its concentration in groundwater will not peak until around 2020, and there will still be significant quantities of the chemical in the water after the middle of the coming century. (*Source: N. L. van der Noot.*)

dichloropropane (DCPa), which, as far as scientists know, has a very long lifetime in groundwater. A calculation for one watershed shows that the DCPa already in the soil will work its way down into groundwater and appear there in significant concentrations only after the year 2000. Thereafter it will contaminate the groundwater for decades in concentrations up to 50 times the European drinking water standard.

A delay in a different sector of World3 is due to the population age structure. A population with a recent history of high birth rates has many more young people than old people. Therefore even if fertility falls, the population keeps growing for decades as the young people reach child-bearing age. Though the number of children per family goes down, the number of families increases. Because of this "population momentum," if the fertility of the entire world population reaches replacement level (about two children per family on average) by the year 2010, the population will go on growing until 2060 and will level off at about 8 billion. If replacement fertility were reached worldwide in 2035, the population would grow until 2095 and reach over 10 billion.[9]

There are many other delays in the "real world" system. Nonrenewable resources may be drawn down for generations before their depletion has serious economic consequences. Industrial capital cannot be built overnight. Once it is in place, it has a lifetime of decades. An oil refinery cannot be converted easily or quickly into a tractor factory or a hospital or a more efficient, less polluting oil refinery.

World3 is replete with delays in its feedback mechanisms, including all those mentioned above. As in the PCB case, we assume a delay between the release of pollution and its noticeable effect on the system. We assume a delay of roughly a generation before couples fully adjust their decisions about family size to changing conditions of income and infant mortality. It takes decades before a new capital plant can be put into place to alleviate a shortage of food or labor or services. It takes time for land fertility to be regenerated or pollution to be absorbed.

We did not include in World3 the time it takes for scientists to understand a problem, or for governments to make decisions, or for material values to change. As it is, the simplest and most incontrovertible physical delays are enough to eliminate smooth sigmoid growth as a possible behavior for the world economic system. Because of the delays in the signals from nature's limits, if there are no self-enforced limits, overshoot is inevitable.

If the warning signal from the limits to the growing entity is delayed, or if the response is delayed, and if the environment is not eroded when overstressed, then the growing entity will overshoot its limit for a while, make a correction, and undershoot, then overshoot again, in a series of oscillations that usually damp down to an equilibrium within the limit (Figure 4-8).

Overshoot and oscillation can occur only if the environment can support the system during periods of overload or repair itself quickly enough to recover during periods of underload.

Renewable resources such as forests, soils, fish, and rechargeable groundwater, are erodable, but they also have a self-regenerating capability. They can recover from a period of overuse, as long as it is not too long and as long as damage to the breeding stock, nutrient source, or aquifer is not too devastating. Given time, soil, seed, and an undisturbed climate, a forest can grow back. A fish population can regenerate. Soils can be restored, especially with active help from farmers. Accumulations of many kinds of pollution can be reduced, if the environment's natural absorption mechanisms have not been badly disturbed.

Therefore the overshoot and oscillation behavior mode is a "real" possibility for the world system. It has been demonstrated in some localities for some resources. New England, for example, has gone through several cycles of building more sawmills than the forest can supply, depleting the harvestable timber, shutting down mills, and then waiting decades until the forest grows back and the overbuilding of sawmills repeats again. The coastal Norwegian fishery has gone through at least one cycle of fish depletion, with the government buying up and retiring fishing boats until the fish stocks could regenerate again.

The down side of an overshoot and oscillation is not a pleasant period to go through. It can mean hard times for industries dependent on an abused resource or bad health in populations exposed to high pollution levels. Oscillations are best avoided. But they are not usually fatal to a system.

However, some overshoots are irreversible. Nothing can bring back an extinct species. Nonrenewable resources such as fossil fuels are permanently destroyed in the very act of using them. Some forms of pollution, such as radioactive materials and toxic heavy metals, can't be rendered harmless by any natural mechanism. Even renewable resources and pollution absorption processes can be permanently eroded by prolonged or systematic misuse. When tropical forests are cut in ways that preclude their regrowth, when rising sea levels infiltrate freshwater aquifers with salt, when soils wash away leaving only bedrock, then the earth's carrying capacity is permanently diminished.

Therefore, the overshoot and oscillation mode is not the only one that could be manifested as humanity approaches the limits to growth. There is one more possibility.

Overshoot and Collapse

If the signal or response from the limit is delayed and if the environment is irreversibly eroded when overstressed, then the growing economy will overshoot its carrying capacity, degrade its resource base, and collapse (Figure 4-8).

The result of this overshoot and collapse is a permanently impoverished environment and a material standard of living much lower than what could have been possible if the environment had never been overstressed.

The difference between the overshoot and oscillation behavior mode and overshoot and collapse is the presence of *erosion loops* in a system. These are positive feedback loops of the worst kind. Normally they are dormant, but when a situation gets bad, they make it worse. They carry a system downward at an ever-increasing pace.

For example, grasslands all over the world have coevolved with grazing animals, from deer to buffalo to antelope to kangaroos. When grasses are eaten down, the remaining stems and roots extract more water and nutrient from the soil and send up more grasses. The number of grazers is held in check by predation on grazing animals and by seasonal migration. The ecosystem does not erode. But if the predators are removed, or the migrations are stymied, or the land is overstocked, an overpopulation of grazers can eat the grass down to the roots.

The less vegetation there is, the less cover there is for the soil. The soil begins to blow away in the wind or wash away in the rain. The less soil, the less vegetation can grow, which allows still more soil to erode away. And so on. Land fertility spirals downward until the grazing range has become a desert.

There are several erosion loops in World3, such as the desertification loop we have just described. Here are some others:

- If people are hungry, they may work the land much more intensively to produce more in the short term at the expense of investment in long-term soil maintenance. Therefore less food leads to lower soil fertility, which brings food down even further.

- When problems appear that require industrial output—pollution abatement equipment, for example, or more agricultural inputs, or equipment for locating and purifying nonrenewable resources—available investment may be allocated to solving the immediate problem, rather than combating depreciation. If the established capital plant begins to deteriorate, that makes even less industrial output available for immediate problems, which may lead to further postponement of capital upkeep, and further erosion in the capital stock.
- If in a weakening economy services per capita go down, family planning expenditures can fail, bringing birth rates up, bringing services per capita down even further.
- If pollution levels build too high, they will pollute the pollution absorption mechanisms themselves, thereby increasing the rate of pollution buildup still more.

This last erosive mechanism, polluting the natural mechanisms of pollution absorption, is particularly insidious, and it is a phenomenon for which we had little evidence when we first postulated it twenty years ago. At the time we had in mind such interactions as dumping pesticides into water bodies, thereby killing the organisms that normally clean up organic wastes; or emitting both nitrogen oxides and volatile organic chemicals into the air, which react with each other to make more damaging photochemical smog.

Since then other examples of the degradation of the planet's own pollution-control devices have come to light. One of them is the apparent ability of short-term air pollutants, such as carbon monoxide, to deplete scavenger hydroxyl radicals in the air. These hydroxyl radicals normally react with and destroy the greenhouse gas methane. When air pollution removes them from the atmosphere, methane concentrations increase. Air pollution can destroy a cleanup mechanism and make global climate change worse.[10]

Another such process is the ability of air pollutants to weaken or kill forests, which then diminishes a major sink for the greenhouse gas carbon dioxide. A third is the effect of acidification—from either fertilizers or acid rain—on soils. At normal levels of acidity, soils are pollution absorbants. They can bind with and sequester toxic heavy metals. But these bonds are broken under acidic conditions.

> As soils acidify, toxic heavy metals, accumulated and stored over long time periods (say, decades to a century) may be mobilized and leached rapidly into ground and surface waters or be taken up by plants. The ongoing acidification of Europe's soils from acid deposition is clearly a source of real concern with respect to heavy metal leaching.[11]

There are more erosive positive feedback loops in the "real world" than the ones we included in World3, most notably social erosion, in which a breakdown in social order feeds upon itself to create further breakdown. It is difficult to quantify erosive mechanisms of any sort. Erosion is a whole-system phenomenon. It has to do with interactions among multiple forces. It appears only at times of stress, and by the time it becomes obvious, there is rarely an opportunity either to study it or to stop it. Whatever its precise nature, however, if there is any possibility that a system contains a latent erosion process, then that system has the possibility, if it is overstressed, of collapse.

On a local scale, overshoot and collapse can be seen in the processes of desertification, mineral or groundwater depletion, poisoning of soils or forests by long-lived toxic wastes. Legions of failed civilizations, abandoned farms, busted boomtowns, and abandoned, toxic industrial lands testify to the "reality" of this system behavior. On a global scale, overshoot and collapse could mean the breakdown of the great supporting cycles of nature that regulate climate, purify air and water, regenerate biomass, preserve biodiversity, and turn wastes into nutrients. Twenty years ago few people would have thought ecological collapse on that scale possible. Now it is the topic of scientific meetings and international negotiations.

World3: Two Possible Scenarios

In the simulated world of World3 the industrial ethic is one of continuous economic growth. The World3 population will stop growing only when it is rich enough. Its resource base is limited and erodable. The feedback loops that connect and inform decisions in the World3 system contain many substantial delays, and the physical processes have considerable momentum. It should come as no surprise that the most common mode of behavior of the model world is overshoot and collapse.

For example, the graphs in Scenario 1 show the behavior of World3 when it is run "as is," with numbers we consider "realistic," and with no unusual technical or policy changes. This computer output can serve as a reference against which to compare scenarios that test policy changes and alternate values for uncertain numbers. Therefore, twenty years ago we called it the "reference run" or "standard run." We did not think it the most probable outcome, and we certainly didn't mean it as a prediction. It is just one of many possibilities. But many people imbued the "standard run" with more importance than the other scenarios that followed. To prevent that from happening this time, we'll give this run another name. Call it Scenario 1.

In Scenario 1 the world society proceeds along its historical path as long as possible without major policy change. Technology advances in agriculture, industry, and social services according to established patterns. There is no extraordinary effort to abate pollution or conserve resources. The simulated world tries to bring all people through the demographic transition and into an industrial and then post-industrial economy. This world acquires widespread health care and birth control as the service sector grows; it applies more agricultural inputs and gets higher yields as the agricultural sector grows; it emits more pollutants and demands more nonrenewable resources as the industrial sector grows.

The global population in Scenario 1 rises from 1.6 billion in the simulated year 1900 to over 5 billion in the simulated year 1990 and over 6 billion in the year 2000. Total industrial output expands by a factor of 20 between 1900 and 1990. Between 1900 and 1990 only 20% of the earth's total stock of nonrenewable resources is used; 80% of these resources remain in 1990. Pollution in that simulated year has just begun to rise noticeably. Average consumer goods per capita in 1990 is at a value of 1968-$260 per person per year—a useful number to remember for comparison in future runs.[12] Life expectancy is increasing, services and goods per capita are increasing, food production is increasing. But major changes are just ahead.

In this scenario the growth of the economy stops and reverses because of a combination of limits. Just after the simulated year 2000 pollution rises high enough to begin to affect seriously the fertility of the land. (This could happen in the "real world" through contamination by heavy metals or persistent chemicals, through climate change, or through increased levels of ultraviolet radiation from a diminished ozone layer.) Land fertility has declined a total of only 5% between 1970 and 2000, but it is degrading at 4.5% per year in 2010 and 12% per year in 2040. At the same time land erosion increases. Total food production begins to fall after 2015. That causes the economy to shift more investment into the agriculture sector to maintain output. But agriculture has to compete for investment with a resource sector that is also beginning to sense some limits.

In 1990 the nonrenewable resources remaining in the ground would have lasted 110 years at the 1990 consumption rates. No serious resource limits were in evidence. But by 2020 the remaining resources constituted only a 30-year supply. Why did this shortage arise so fast? Because exponential growth increases consumption and lowers resources. Between 1990 and 2020 population increases by 50% and industrial output grows by 85%. The nonrenewable resource use rate doubles. During the first two decades of the simulated twenty-first century, the rising population and industrial plant in Scenario 1 use as many nonrenewable resources as the global economy used in the entire century before. So many resources are used that much more capital and energy are required to find, extract, and refine what remains.

As both food and nonrenewable resources become harder to obtain in this simulated world, capital is diverted to producing more of them. That leaves less output to be invested in basic capital growth.

Scenario 1 THE "STANDARD RUN" FROM *THE LIMITS TO GROWTH*

The world society proceeds along its historical path as long as possible without major policy change. Population and industry output grow until a combination of environmental and natural resource constraints eliminate the capacity of the capital sector to sustain investment. Industrial capital begins to depreciate faster than the new investment can rebuild it. As it falls, food and health services also fall, decreasing life expectancy and raising the death rate.

Scenario 1

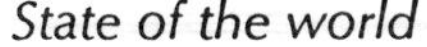

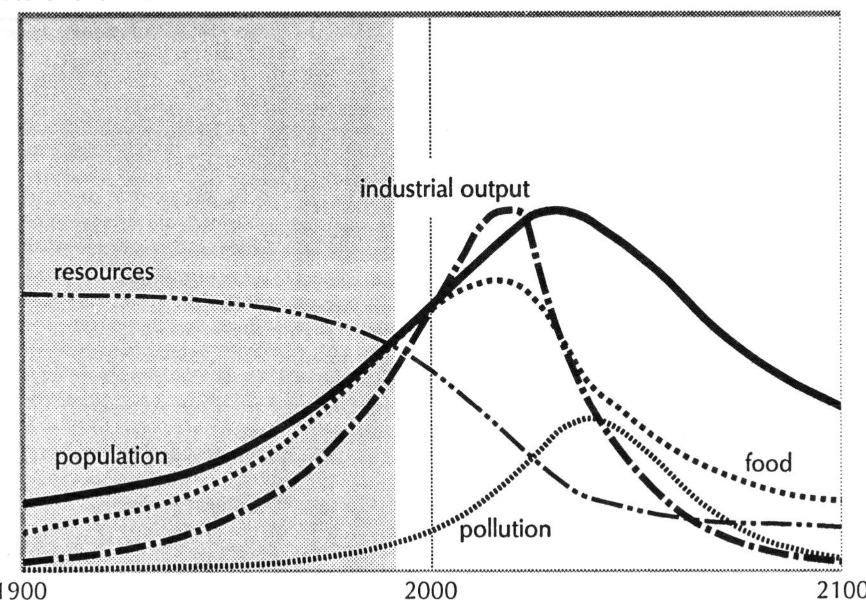

Material standard of living

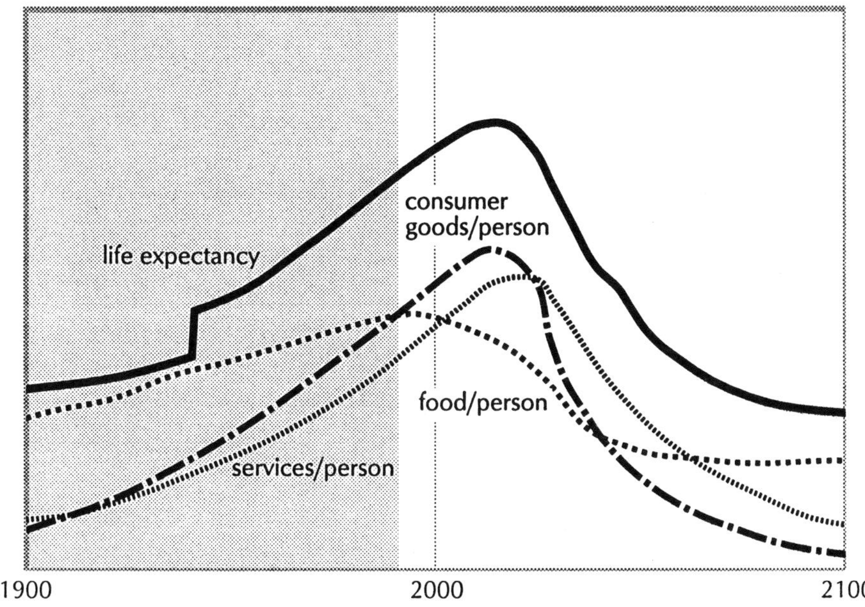

Finally investment cannot keep up with depreciation (this is *physical* investment and depreciation, not monetary). The economy cannot stop putting its capital into the agriculture and resource sectors; if it did the scarcity of food, materials, and fuels would restrict production still more. So the industrial capital plant begins to decline, taking with it the service and agricultural sectors, which have become dependent upon industrial inputs. For a short time the situation is especially serious, because the population keeps rising, due to the lags inherent in the age structure and in the process of social adjustment. Finally population too begins to decrease, as the death rate is driven upward by lack of food and health services.

This scenario is *not a prediction.* It is not meant to forecast precise values of any of the model variables in the future, nor the exact timing of events, nor, we believe, does it necessarily represent the most likely "real world" outcome. (We'll show another possibility in a moment, and many more in Chapters 6 and 7.) The strongest statement of certainty we can make about Scenario 1 is that it portrays the most likely *general behavior mode* of the system, *if* the policies that influence economic growth and population growth in the future are similar to those in the past, *if* technologies and value changes continue to evolve in the manner prevailing now, and *if* the uncertain numbers in the model are roughly correct.

What if they aren't correct? Since many of the critical numbers (such as the amount of nonrenewable resources still to be discovered) are simply unknowable, the model can be used to test the range of uncertainty, to see what difference it would make if, for example, there are twice as many nonrenewable resources waiting to be discovered under the ground as we assumed. That test is shown in Scenario 2. As

Scenario 2 DOUBLED RESOURCES ARE ADDED TO SCENARIO 1

If we double the natural resource endowment we assumed in Scenario 1, industry can grow 20 years longer. Population rises to more than 9 billion in 2040. These increased levels generate much more pollution, which reduces land yield and forces much greater investment in agriculture. Eventually declining food raises the population death rate.

SCENARIO 2

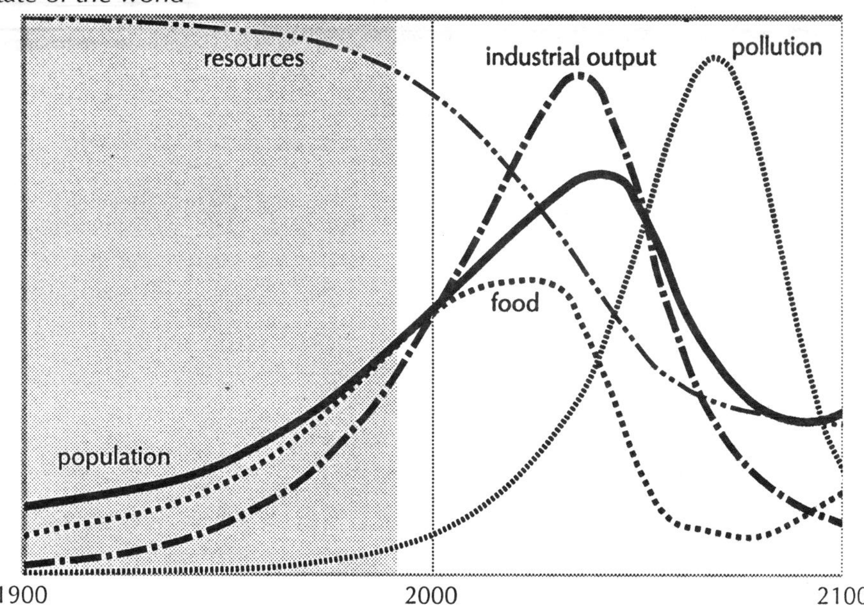
State of the world
resources
industrial output
pollution
food
population
1900
2000
2100

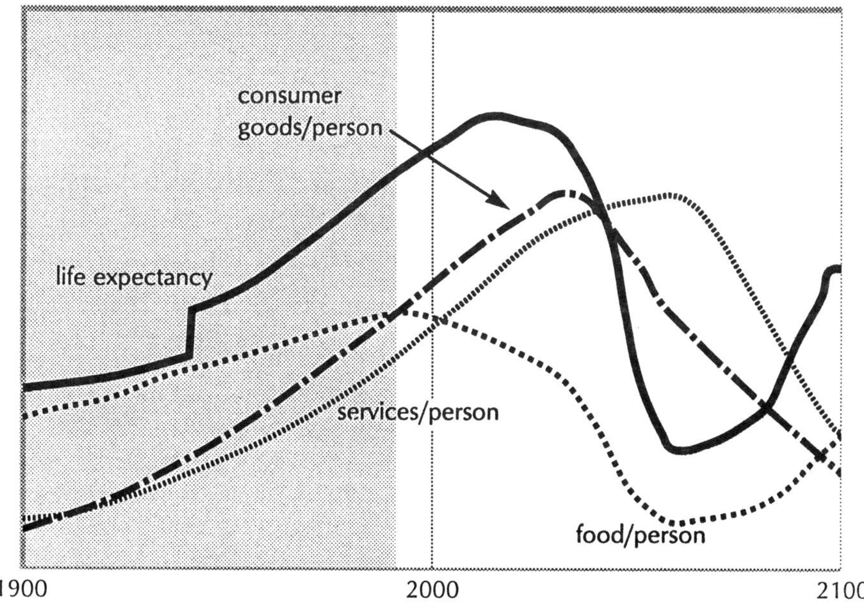
Material standard of living
consumer
goods/person
life expectancy
services/person
food/person
1900
2000
2100

you can see, resources last considerably longer in this simulation than they did in Scenario 1. But the general behavior of the model is still overshoot and collapse. This time the collapse comes for a different combination of reasons.

The additional resources allow industry to grow twenty years longer than it did in Scenario 1. Population also grows longer, reaching a total of almost 9.5 billion in the simulated year 2040. Greater rates of industrial output cause pollution to grow more; pollution in Scenario 2 peaks about thirty years later than it does in Scenario 1, and at a level more than 3 times higher. Part of its rise is due to greater pollution generation rates, and part is due to the fact that the simulated world's pollution assimilation processes are becoming impaired. At the pollution peak in 2070 the average lifetime of pollutants in the environment has more than doubled.

The pollution has a major impact on land fertility, which declines dramatically. Investments in agriculture increase to combat that loss, but food production still falls sharply. Death rates rise from lack of food. Capital growth stops as investment is pulled into agriculture and eventually into a depleted nonrenewable resource sector as well.

Which is a more likely future, Scenario 1 or Scenario 2? If there were a scientific way of answering that question, it would depend on evidence about the "actual" amount of undiscovered nonrenewable resources. But in fact the question of whether Scenario 1 or Scenario 2 is most "realistic" is not worth debating. Neither is "realistic." Neither can foretell a future in which human beings can change their reactions to events, can foresee events, and can change the structure of their system. There are many more uncertain numbers to test, and many technical and social policies to try. We'll come to them in Chapters 6 and 7. All that World3 has told us so far is that the model system, and by implication the "real world" system, has a strong tendency to overshoot and collapse. In fact, in the thousands of model runs we have tried over the years, overshoot and collapse has been by far the most frequent outcome. By now the reasons for that should be quite clear.

Why Overshoot and Collapse?

A population and economy are in overshoot mode when they are drawing resources or emitting pollutants at an unsustainable rate, but the stresses on the support system are not yet strong enough to reduce the rates of withdrawal or emission. Overshoot comes from delays in feedback—from the fact that decision makers in the system do not get, or believe, or act upon information that limits have been exceeded until long after they have been exceeded.

Overshoot is only possible because there are accumulated resource stocks that can be drawn down. One can cut a forest beyond its annual growth rate for quite a long time, because there is a standing stock of wood in the forest that has grown over decades or centuries. One can build up enough herds to overgraze, or boats to overfish, because there are accumulations of forage and fish that were not exploited in the past. The larger the accumulated stocks, the higher and longer the overshoot can be. If a society takes its signals from the simple availability of stocks, rather than from their size, quality, diversity, health, and rates of replenishment, it will overshoot.

Physical momentum causes delay not in the warning signals, but in the response to the signals. Because of the time it takes forests to grow, populations to age, pollutants to work their way through the ecosystem, polluted waters to clear, capital plants to depreciate, and people to be educated or retrained, the economic system can't change overnight, even if it gets and acknowledges clear and timely signals that it should do so. To steer correctly, a system with inherent physical momentum needs to be looking decades ahead.

The final contributor to overshoot is growth. If you're driving a car with fogged windows or faulty brakes, the first thing you would do to avoid overshoot would be to *slow down.* You would certainly not insist on accelerating. Delays in feedback can be handled, as long as the system is not moving too fast to receive and respond to one signal before the next signal comes in. Constant acceleration will take any system to the point where it can't respond in time. Even a car and driver functioning perfectly are unsafe at high speeds. The faster the growth, the higher the overshoot, and the farther the fall.

What finally converts overshoot to collapse is erosion, aided by nonlinearities. Erosion is a stress to the system that multiplies itself, if it is not quickly remedied. Nonlinearities like the ones shown in Figures 4-1 and 4-6 are equivalent to *thresholds,* beyond which a system's behavior suddenly changes. A population's food supply can be decreased with no impacts on health for a long time, but if food per capita gets below a certain limit, death rates rise sharply. A nation can mine copper ore down to lower and lower grades, but below a certain grade mining costs rise greatly. Soils can erode with no effect on crop yields until that point where the soil becomes more shallow than the root zone of the crop. The presence of thresholds makes the consequences of feedback delays even more serious. If you're driving that car with the fogged windows and faulty brakes, sharp curves in the road mean you need to go even more slowly.

Any population-economy-environment system that has feedback delays and slow physical responses, that has thresholds and erosive mechanisms, is literally *unmanageable.* No matter how brilliant its technologies, no matter how efficient its economy, no matter how wise its decision makers, it simply can't steer itself away from hazards unless it tests its limits very, very slowly. If it constantly tries to accelerate, it is bound to overshoot.

By definition overshoot is a condition in which the delayed signals from the environment aren't yet strong enough to force an end to growth. How, then, can a society tell if it is in overshoot? Falling resource stocks and rising pollution sinks are the first clues. Here are some others:

- Capital, resources, and labor must be diverted from final goods production to exploitation of more scarce, more distant, deeper, or more dilute resources.

- Capital, resources, and labor must be diverted from final goods production to activities that compensate for what used to be free services from nature (for example, sewage treatment, air purification, flood control, pest control, restoration of soil nutrients, pollination, or the preservation of species).

- Capital, resources, and labor are used to protect, defend, or gain access to resources that are increasingly concentrated in just a few remaining places.
- Natural pollution-cleanup mechanisms begin to fail.
- Capital depreciation is allowed to exceed investment, or maintenance is deferred, so there is deterioration in capital stocks, especially in long-lived infrastructure.
- Investment in human resources (education, health care, shelter) is decreased in order to meet immediate consumption needs or to pay debts.
- Debts become a higher percentage of annual real output.
- Conflicts increase, especially conflicts over sources or sinks. There is less social solidarity, more hoarding, greater gaps between haves and have-nots.

A period of overshoot does not necessarily lead to collapse. It does call for fast and determined action, however, if collapse is to be avoided. The resource base must be protected quickly. The drains on it must be sharply reduced. That need not mean reducing population or capital or living standards, though it certainly means reducing their growth wherever possible. What must go down quickly, are material and energy throughputs. Fortunately (in a perverse way) there is so much waste and inefficiency in the current global economy, that there is tremendous potential for reducing throughputs while still raising the quality of life. And then the next task is to restructure the system so that overshoot never happens again.

In summary, here are the central assumptions in the World3 model, which give it a strong tendency to overshoot and collapse. You can decide for yourself whether they are also characteristic of the "real world."

- Growth is inherent to the human value system, and growth of both the population and the economy, when it does occur, is exponential.

- There are physical limits to the sources of materials and energy that sustain the human population and economy, and there are limits to the sinks that absorb the waste products of human activity.
- The growing population and economy receive signals about physical limits that are distorted and delayed. The response to those signals is also delayed.
- The system's limits are not only finite, but erodable when they are overstressed or overused.

If those are the causes of overshoot and collapse, then they are also the keys to avoiding that behavior. To change the system, to make it manageable and sustainable, the same structural features can be reversed:

- Throughputs of energy and materials can be reduced by increasing their efficiency.
- Limits can be raised as far as possible by affordable technologies.
- Signals can be improved and reactions speeded up; society can look further ahead when it evaluates the costs and benefits of current choices.
- Erosion can be prevented and, where it already exists, reversed.
- Growth of population and capital can be slowed and eventually stopped.

In Chapters 6 and 7 we will show the effect of these changes in the World3 system. But first a short digression for a story that illustrates all the dynamic principles we have presented in this chapter.

chapter 5

Back from Beyond the Limits: The Ozone Story

We find ourselves, one way or another, in the midst of a large-scale experiment to change the chemical construction of the stratosphere, even though we have no clear idea of what the biological or meteorological consequences may be.

F. Sherwood Rowland[1]

The human race has recently overshot, learned about, and backed off from one clear environmental limit—the destruction of the stratospheric ozone layer. The ozone story is a hopeful one, so far at least. It shows the people and nations of the world at their collective best, though it also demonstrates some common human failings.

Scientists sounded the first warnings about the disappearing ozone layer and then transcended political boundaries to form an impressive knowledge-gathering force. But they could do that only after they managed to get beyond their own perceptual blinders. Governments and corporations at first acted as doubters and foot-draggers, but then some of them emerged as true leaders. Environmentalists were labeled as wild-eyed alarmists, but in this case they turned out to have underestimated the problem.

The United Nations in this story showed its potential for passing crucial information around the world and for providing neutral ground and sophisticated facilitation as governments worked through an undeniably international problem. Third World nations found in the ozone crisis a new power to act on their own behalf, by refusing to cooperate until they were guaranteed technical and financial support for that cooperation.

In the end, the world's nations acknowledged that they had overrun a serious limit. Soberly, reluctantly, they agreed to give up a profitable and useful industrial product. They did it before there was any measurable economic, ecological, or human damage and before there was complete scientific certainty. They may have done it in time.

The Growth

Chlorofluorocarbons (CFCs) are some of the most useful compounds ever invented by human beings (see Table 5-1). They are nontoxic and stable. They do not burn or react with other substances or corrode materials. Because they have low thermal conductivity, they make excellent insulators when blown into plastic foam for hot-drink cups, hamburger containers, or wall insulation. Some CFCs evaporate and recondense at room temperatures, which makes them perfect coolants for refrigerators and air conditioners. (In that use they are known under the trade name Freon.) CFCs make good solvents for cleaning metals, from the intricate microspaces on electronic circuit boards to the rivets that hold together airplanes. CFCs are inexpensive to make, and they can be discarded safely—or so everyone thought—simply by releasing them as gases into the atmosphere.

As Figure 5-1 shows, from 1950 to 1975 world production of CFCs grew at 7% to 10% per year—doubling every 10 years or less. By the 1980s the world was manufacturing a million tonnes of CFCs annually. In the United States alone CFC coolants were at work in 100 million refrigerators, 30 million freezers, 45 million home air conditioners, 90 million car air conditioners, and hundreds of thousands of coolers in restaurants, supermarkets, and refrigerated trucks.[2] The average North American or European was using 2 pounds (0.85 kg) of CFCs per year. The average resident of China or India was using less than an ounce (0.03 kg)[3]. For an increasing number of chemical companies in North

Table 5-1 USES, PRODUCTION RATES, AND RESIDENCE TIMES OF THE IMPORTANT OZONE-DEPLETING CHEMICALS

Compound name	*Chemical formula*	*Ozone depletion potential*	*Uses*	*1985 world production (tonnes)*	*Residence time in atmosphere (years)*
CFC-011	$CFCl_3$	1.0	refrigeration, aerosol, foam	298,000	65–75
CFC-012	CF_2Cl_2	0.9–1.0	refrigeration, aerosol, foam, sterilization, food freezing, heat detectors, warning devices, cosmetics, pressurized blowers	438,000	100–140
CFC-113	CCl_3CF_3	0.8–0.9	solvent, cosmetics	138,500	100–134
CFC-114	$CClF_2CClF_2$	0.7–1.0	refrigeration		300
CFC-115	$CClF_2CF_3$	0.4–0.6	refrigeration, whipped topping stabilizer		500
Halon 1301	$CBrF_3$	10.0–13.2	fire fighting	2,600	110
Halon 1211	$CClBrF_2$	2.2–3.0	fire fighting	2,600	15
HCFC-22	$CHClF_2$	0.05	refrigeration, aerosol, foam, fire fighting	81,200	16–20
Methyl chloroform	CH_3CCl_3	0.15	solvent	499,500	5.5–10
Carbon tetrachloride	CCl_4	1.2	solvent	71,200	50–69

Figure 5-1 World Reported Production of CFC-011 and CFC-012

Thousands of tonnes per year

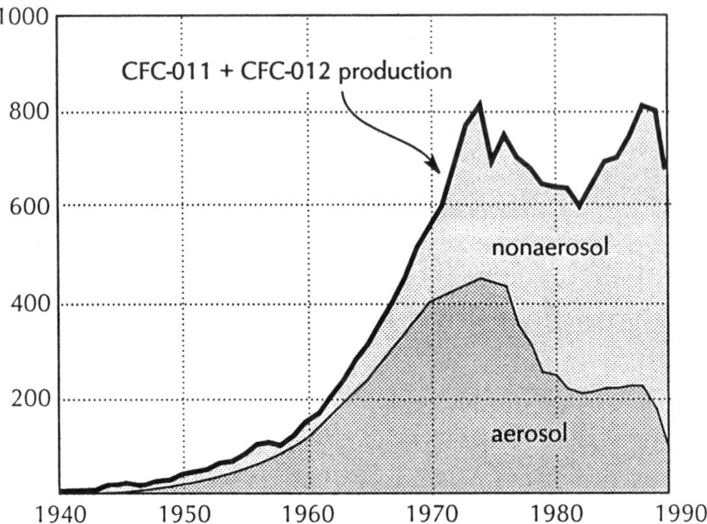

Production of the two most widely used CFCs grew rapidly until 1974, when the first papers postulating their effects on the ozone layer appeared. The subsequent decrease was due to environmental activism against CFC-containing aerosol spray cans, which were finally banned in the United States in 1978. After 1982 the expansion of other CFC uses caused total production to rise again. (*Source: Chemical Manufacturers Association.*)

America, Europe, the Soviet Union, and Asia these substances were a major source of income. For even more companies they were necessary to the production process.

The Limit

High up in the stratosphere, twice as high as Mount Everest or as jet planes fly, is a gossamer veil with a crucial function. It is made of the gas called ozone—three oxygen atoms stuck together (O_3), as opposed to the ordinary oxygen gas of the atmosphere, which is two oxygen atoms stuck together (O_2). Ozone is unstable; it is so reactive that it attacks and oxidizes almost anything it contacts. Therefore, in the lower atmosphere, which is dense with materials with which it can react, including plant tissue and human lungs, ozone is a destructive but short-lived air pollutant. In the stratosphere, however, there isn't much for an ozone molecule to run into. Ozone is constantly created there by the action of sunlight on ordinary oxygen molecules, and it lasts a relatively long time. Therefore an "ozone layer" accumulates.

The ozone layer is rich in ozone only in comparison to the scarcity of that gas elsewhere in the atmosphere. Only one molecule in 100,000 in the ozone layer is actually ozone. But there is enough ozone in the stratosphere to absorb from the sun's incoming light most of a particularly harmful ultraviolet wavelength called UV-B (see Figure 5-2). UV-B light is a stream of little bullets of energy of just the right frequency to take apart organic molecules—the kinds of molecules that make up all life, including the DNA molecules that carry the code for life's reproduction.

When living organisms are hit by UV-B energy bullets, one possible result is cancer. UV-B light has long been known to cause skin cancer in laboratory animals. Nearly all human skin cancers occur on body parts exposed to the sun. They occur especially in fair-skinned people who spend considerable time in the sun. Australia has the highest rate of skin cancer in the world: at current rates of incidence, two of every three Australians will develop some kind of skin cancer during their lifetimes, and 1 in 60 will develop the most deadly type, malignant melanoma. Scientists estimate that for every 1% decrease in the ozone layer, there will be an increase of 2% in UV-B radiation at the earth's surface, and an increase of 3% to 6% in the incidence of human skin cancer.[4]

UV-B radiation puts the human skin in double jeopardy. It can induce the growth of a cancer, and it can also suppress the immune system's ability to fight cancer. This suppression of the immune system also makes people more susceptible to herpes and other infectious diseases.

Beside the skin, the other part of the body most exposed to UV-B radiation is the eye. Ultraviolet light can burn the cornea, causing a condition known as "snow blindness," because it often afflicts skiers and mountaineers at high altitudes. Occasional snow blindness is very

Figure 5-2 Absorption of Ultraviolet Light by the Atmosphere

Solar energy (watts per square meter)

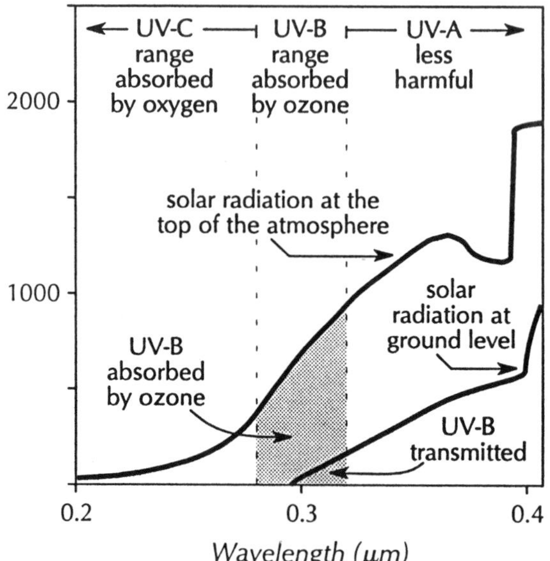

Incoming ultraviolet light from the sun is almost totally absorbed by oxygen and ozone in the atmosphere. (*Source*: *United Nations Environmental Programme.*)

painful; repeated snow blindness can reduce vision permanently. UV-B light can also damage the retina and generate cataracts in the eye's lens.

If ozone depletion allows more UV-B light to reach the earth's surface, any animal with eyes and skin exposed to the sun would be expected to suffer effects similar to those in human beings. Detailed studies of the effects of UV-B on other animals and plants are just beginning, but some results are already clear:

- Single-celled and very small organisms are more likely to be damaged than large organisms because UV-B light can only penetrate a few layers of cells.
- UV-B light enters only the top few meters of the ocean, but this is the layer where most aquatic microorganisms live. These small,

floating plants and animals are particularly sensitive to UV-B radiation. They are also the base of most ocean food chains. Therefore an increase in UV-B could greatly perturb many populations of ocean life.

- Exposure to UV-B light decreases leaf area, plant height, and photosynthesis in green plants. Different agricultural crops respond to UV-B radiation to different extents, but in two-thirds of the crop plants studied, yields go down as UV-B goes up. For example, studies of UV-B light on soybeans lead to the conclusion that each 1% depletion in the ozone layer will result in a 1% decline in soybean yield.[5]
- Cultivated plants seem more sensitive to UV-B light than weeds.

Living creatures have evolved many ways to protect themselves from ultraviolet light, such as pigmentation, coverings of hair or scales, mechanisms to repair damaged DNA, and behavior patterns that keep sensitive creatures hidden from strong sunshine. These devices work better for some species than others. Therefore one effect of a degraded ozone layer would be population decreases or extinctions in some species and population increases in others. Grazers could grow out of balance with their forage supply, pests with their predators, parasites with their hosts. Every ecosystem would feel the effect of a diminished ozone layer in ways that are impossible to predict, especially if other changes, such as climate changes, are going on at the same time.

The First Signals

In 1974 two scientific papers were published independently, both of which suggested a threat to the ozone layer. One said that chlorine atoms in the stratosphere could be powerful ozone destroyers.[6] The second said that CFCs were reaching the stratosphere and breaking up, releasing chlorine atoms.[7] Taken together, these publications predicted that human CFC use could trigger a hitherto unsuspected environmental disaster.

Because they are unreactive and insoluble, CFCs do not dissolve in rain or react with other gases. Their carbon-chlorine and carbon-fluorine bonds are so strong that the wavelengths of sunlight that reach the lower atmosphere do not break them. About the only way a CFC molecule can be cleansed from the atmosphere is to rise very high, above most of the air of the planet, to where it finds the short-wavelength ultraviolet light that never reaches the earth's surface because ozone and oxygen filter it out. That radiation eventually breaks up the CFC molecule, releasing free chlorine atoms.

That's where the trouble begins. Free chlorine (Cl) can react with ozone to make oxygen and chlorine oxide (ClO). Then the ClO reacts with an oxygen atom (O) to make O_2 and *Cl again*. The Cl atom can then turn another ozone molecule into oxygen and be regenerated yet again (Figure 5-3).

One Cl atom can cycle through this series of reactions over and over, destroying one ozone molecule each time. Chlorine acts like a Pac-Man of the high atmosphere, gobbling one ozone molecule after another and then being regenerated to gobble again. The average Cl atom can destroy about 100,000 ozone molecules before it is finally removed from the stratosphere. In the usual path of removal Cl reacts with methane to produce hydrochloric acid (HCl). At that point two things can happen; either the HCl can break up, release Cl again, and continue the cycle of ozone destruction, or the HCl can sink down into the lower atmosphere, where it typically dissolves in water and comes back to earth as acid rain.

The continuous chemical regeneration of Cl is only one insidious characteristic of the ozone breakdown process. Another is the long delay between the human synthesis of a CFC molecule and its arrival in the stratosphere. For some uses (such as aerosol propellants) production is followed quickly by discharge into the air. For other uses (such as refrigerants and foam insulation) the CFC may be released years or even decades after its production. After release it takes about *fifteen years* for a CFC molecule released on the earth's surface to work its way up to the high stratosphere where it breaks down and reacts with ozone. So the thinning of the ozone layer measured at any time is a result of CFCs manufactured and released fifteen or more years ago.

Figure 5-3 How CFCs Destroy Stratospheric Ozone

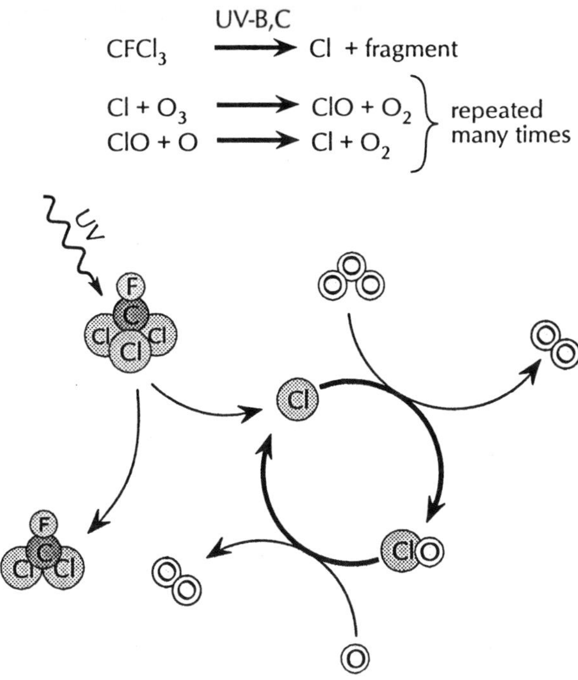

CFC molecules high in the stratosphere are broken up by ultraviolet light to release free chlorine atoms (Cl). These atoms react with ozone (O_3) to produce chlorine oxide (ClO). The ClO then can react with a oxygen atom to release Cl again, which can react with another ozone molecule—and so on.

The First Response

The two 1974 papers predicting the erosion of the ozone layer started a burst of research on atmospheric chlorine chemistry around the world. In the United States the scientific information also made its way quickly into the political process. That happened partly because the authors of the first papers were American, worried about their findings, and energetic in bringing them to public attention (especially F. Sherwood Rowland, who brought the matter quickly to the National Academy of Sciences and to the Congress). Another factor that created political discussion rapidly in the United States was the large, well-organized environmental movement.

When American environmentalists understood the implications of the CFC-ozone connection, they went into action. They started by condemning the use of aerosol spray cans. It is crazy, they said, to threaten life on earth just for the privilege of spraying on your deodorant. Their choice of aerosol cans as a target was oversimplified, since non-CFC aerosol propellants were also in use, and since there were many other uses of CFCs. But to keep things simple aerosol cans were branded, and consumers responded. Sales of aerosol cans plummeted by over 60%. Political pressure mounted for a law to ban CFC-containing aerosols entirely.

There was, as you might expect, industry resistance to this development. A Du Pont executive testified before Congress in 1974 that "The chlorine-ozone hypothesis is at this time purely speculative with no concrete evidence to support it." But he said, "If creditable scientific data . . . show that any chlorofluorocarbons cannot be used without a threat to health, Du Pont will stop production of these compounds."[8] Fourteen years later Du Pont, the world's largest producer of CFCs, honored that pledge.

A law forbidding the use of CFCs as aerosol propellants was passed in the United States in 1978. Together with the consumer action that had already reduced aerosol sales, that ban produced a 25% drop in worldwide manufacture of CFCs. In most of the rest of the world, however, aerosol sprays still contained CFCs, and other uses of CFCs, especially in the electronics industry, continued to climb. By the mid-1980s worldwide CFC use was back up to its 1975 peak (Figure 5-1).

Erosion: The Ozone Hole

In October 1984 scientists of the British Antarctic Survey measured a 40% decrease in ozone in the stratosphere over their survey site at Halley Bay in Antarctica. Their ozone measurements had been declining steadily for about ten years (Figure 5-4). But the scientists had been reluctant to believe what they were seeing. A 40% drop seemed impossible. Computer models based on knowledge of atmospheric chemistry

Figure 5-4 OZONE MEASUREMENTS AT HALLEY BAY, ANTARCTICA

Total ozone concentration (Dobson units)

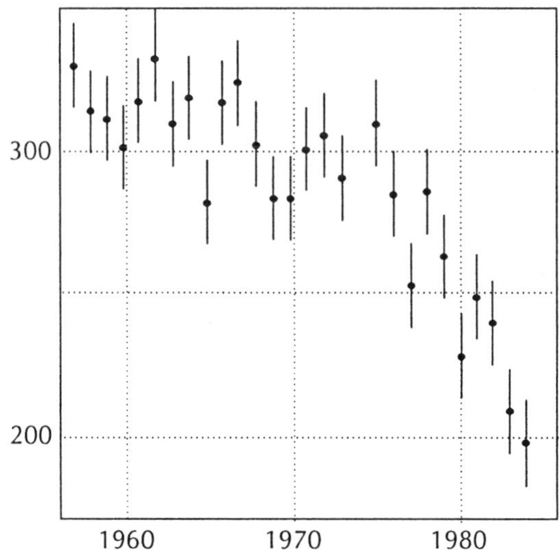

Ozone concentrations in the atmosphere above Halley Bay, Antarctica, during the month of October, taken as the sun returned each southern spring, had been declining for more than a decade before the paper announcing the ozone hole was published in 1985. (*Source*: *J. C. Farman et al.*)

at the time were predicting only a few percent decline in ozone, at most.

The British scientists rechecked their instruments. They looked for confirming measurements from some other part of the earth. Finally they found one. A second measuring station about 1600 kilometers (1000 miles) to the northwest also reported enormous decreases in stratospheric ozone.

In May 1985 the historic paper was published that announced an "ozone hole" in the Southern Hemisphere.[9] The news reverberated around the scientific world. Scientists at the National Aeronautics and Space Administration (NASA) of the United States scrambled to check readings on atmospheric ozone made by the Nimbus 7 satellite, measurements that had been taken routinely since 1978. Nimbus 7 had never indicated an ozone hole.

Figure 5-5 AS REACTIVE CHLORINE INCREASES, ANTARCTIC OZONE DECREASES

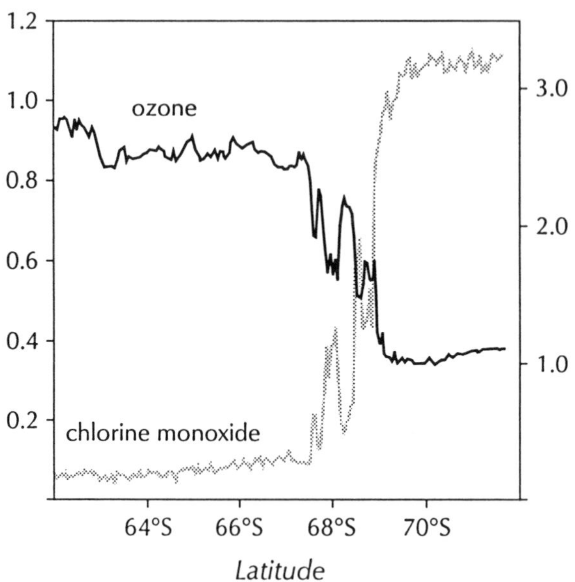

Instruments aboard NASA's ER-2 research airplane measured concentrations of chlorine monoxide and ozone simultaneously as the plane flew from Punta Arenas, Chile (53°S) to 72°S. The data shown above were collected on September 16, 1987. As the plane entered the ozone hole, the concentration of chlorine monoxide increased to about 500 times normal levels, while the ozone concentration plummeted. (*Source: J. G. Anderson et al.*)

Checking back, NASA scientists found that their computers had been programmed to reject very low ozone readings on the assumption that such low readings must indicate instrument error.[10] Fortunately the measurements thrown out by the computer were recoverable. They confirmed the Halley Bay observations. They showed that ozone levels had been dropping over the South Pole for a decade. Furthermore they provided a detailed map of the hole in the ozone layer. It was enormous, about the size of the continental United States, and it had been getting larger and deeper every year.

Why a hole? Why over Antarctica? What did this finding portend for the entire earth's protection from UV-B radiation? The work of scientists over the next few years to solve this mystery was extraordinary. One of the most spectacular pieces of evidence that chlorine was indeed the culprit causing the ozone hole was gathered in September 1987, when scientists flew an airplane from South America directly toward the South Pole and into the ozone hole. Their measurements of ozone and ClO as they flew are shown in Figure 5-5. Rises and drops in ozone are almost exactly mirrored by drops and rises in ClO.[11] Furthermore, the measured ClO concentrations in the "hole" are hundreds of times higher than any level that could be explained by normal atmospheric chemistry. This figure is often referred to as the "smoking gun" that proved even to the CFC manufacturers that the ozone hole is not a normal phenomenon. It is a sign of a highly perturbed atmosphere, caused by human-produced chlorine-containing pollutants.

It took several years for scientists to come up with an explanation for the hole. In a nutshell, here it is.

Since Antarctica is surrounded by oceans, winds can circle around the continent uninterrupted by land masses. In the southern winter, they create a *circumpolar vortex,* a whirl of winds that traps air over Antarctica and keeps it from mixing with the rest of the atmosphere. The vortex sets up an isolated "reaction vessel" of polar atmospheric chemicals. (There is not such a strong vortex around the North Pole, so the northern ozone hole is much less pronounced.)

In winter the Antarctic stratosphere is the coldest place on earth (down to −90 degrees C). In that extreme cold water vapor hovers as a fog of minute ice crystals high up where the ozone layer is. The surfaces of these innumerable tiny crystals enhance the chemical reactions that release ozone-destroying chlorine.

The chlorine atoms formed in the dark of the Antarctic winter do not immediately enter the chain reaction of ozone destruction. Instead they react just once with ozone to form ClO. The ClO molecules come together to form a relatively stable ClOOCl dimer. An accumulation of ClOOCl builds up, poised and waiting for the return of the sun.[12]

When the light returns in the Antarctic spring, solar radiation breaks up the ClOOCl dimer to release an enormous burst of Cl, which goes to work on the ozone. Ozone concentration drops precipitously within a few weeks. At some altitudes more than 97% of the ozone vanishes.

The returning sunlight gradually dissipates the circumpolar vortex, allowing south polar air to mix again. Ozone-depleted air is dispersed over the rest of the globe, as ozone levels over Antarctica return nearly to normal.

Lesser holes have been observed over the North Pole in the northern spring. Discrete holes are not expected to be found elsewhere. But as the gases in the atmosphere mix, the concentration of ozone in the stratosphere above the whole earth is decreasing measurably. Because of the long delays in CFCs reaching the stratosphere, more ozone depletion is inevitable. Because of the long lifetimes of CFCs and Cl in the atmosphere, the depletion will last for at least a century, even if all CFC releases stop immediately.

The Next Response

There is some disagreement among the people who were involved in the global negotiations about whether the announcement of the ozone hole in 1985 energized politicians as thoroughly as it did scientists. International discussions were already underway to limit CFC production, but they had not made much progress. A meeting in Vienna held two months before the published announcement of the ozone hole produced a feel-good statement that nations should take "appropriate measures" to protect the ozone layer, but it set no timetables and stipulated no sanctions. Industry had abandoned its search for CFC substitutes, since it was not apparent that they would be needed any time soon.[13] The Antarctic ozone hole had not at that time been definitely linked to CFCs; it would not be until three years later.

Something happened, however, between March 1985 in Vienna when there was no real action, and October 1987 in Montreal, when the first international ozone-protection protocol was signed. The hole over the Antarctic did have a psychological effect, maybe all the more so because it was not understood. There was no doubt that the ozone layer was doing strange things. Though there was as yet no proof, CFCs were the most likely culprits.

Proof or no proof, probably nothing would have happened, if it had not been for the United Nations Environment Program (UNEP). UNEP hosted and prodded the international political process. Its staff assembled and interpreted the scientific evidence, presented it to governments, created a neutral venue for high-level discussions, and acted as mediators. UNEP's director, Mustafa Tolba, proved a skilled environmental diplomat, remaining neutral in the many squabbles that arose, patiently reminding everyone that no short-term or selfish consideration was as important as the integrity of the ozone layer.

The negotiating process was not easy.[14] The world's nations had never before confronted a global environmental problem before it was completely understood and before it had produced any measurable damage to human health or to the economy. Major CFC-producing nations played predictable roles in trying to block any strong cutback in CFC use.

Critical decisions sometimes hung on delicate political threads. The United States, for example played a strong leadership role, which was several times nearly undercut by deep divisions within the Reagan administration. Those divisions came to public attention when Interior Secretary Donald Hodel was quoted as saying that the ozone layer would be no problem, if people would just wear broad-brimmed hats and sunglasses when they went outside. The international ridicule that was heaped upon that statement helped those members of the U.S. administration who were trying to get the president to take the ozone problem seriously.

UNEP pressed on. Environmental groups in Europe and the United States put heat on their governments. Scientists conducted workshops to educate journalists, parliamentarians, and the public. Responding to pressures from all sides, national governments finally–and surprisingly quickly–signed in Montreal in 1987 a Protocol on Substances That Deplete the Ozone Layer. The "Montreal Protocol" stipulated first that world production of the five most commonly used CFCs should be frozen at 1986 levels. Then production should be reduced by 20% by 1993, and finally by another 30% by 1998. This "freeze-20-30" agreement was signed by 36 nations, including all the major producers of CFCs.

Figure 5-6 REAL AND PROJECTED GROWTH OF STRATOSPHERIC INORGANIC CHLORINE CONCENTRATIONS DUE TO CFC EMISSIONS

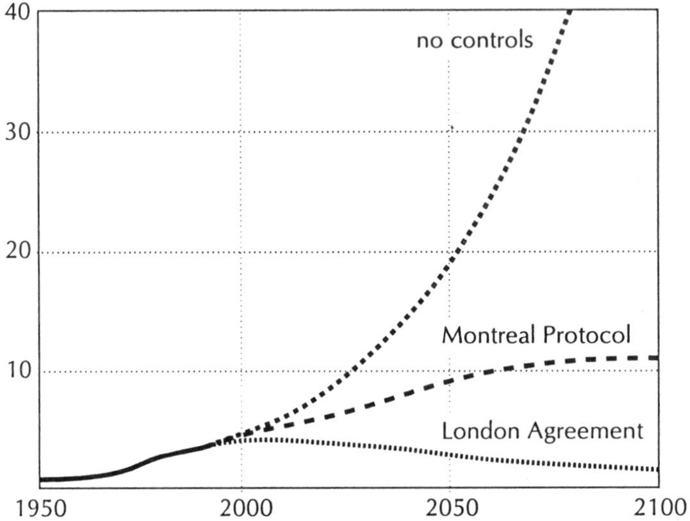

Sustaining the 1986 production rate of CFCs would have led stratospheric chlorine concentrations to increase more than 60-fold between 1950 and 2100. The Montreal Protocol defined lower emission rates, but it would still have permitted chlorine levels to nearly double from their 1980 levels. The London Agreement phases out CFC use; it will lead to declining chlorine levels beginning around the year 2005. (*Sources: J. Hoffman et al.; R. E. Benedick.*)

The Montreal Protocol was a historic agreement. It went far beyond what environmentalists at the time thought was politically possible. And it soon became apparent that the CFC reductions it had called for were not enough. Figure 5-6 shows what would have happened to the concentration of ozone-destroying Cl in the stratosphere, if emissions had continued at the 1986 rate and if they were cut according to the Montreal Protocol. Because of the large stocks of CFCs that have been produced but not yet released, and that have been released but have not yet reached the stratosphere, Cl would have gone on increasing in either case. Even with the Protocol, stratospheric Cl would eventually double its current level.

The reasons for the weakness of the agreement were understandable. Most Third World countries did not sign. China, for example, was trying to equip most of its households with their first refrigerators—that meant a huge new demand for Freon. The USSR waffled, saying that its five-year planning process did not allow rapid change in CFC production. It demanded and got a slower phase-down schedule. And most industrial makers of CFCs were still hoping to maintain at least part of their market.

Within a year after the Montreal Protocol was signed, however, even greater ozone depletion levels were measured, and the "smoking gun" evidence was published. At that point Du Pont announced that it would phase out its manufacture of CFCs completely. In 1989 the United States and the nations of the European Community decided that they would stop all production of the five most common CFCs by the year 2000. They called upon the world to invoke the stipulations that had been written into the Montreal document requiring periodic reassessment of the ozone situation and stronger measures, if necessary.

After further negotiations, again led by UNEP, governments from 92 countries met in London in 1990 and agreed to phase out all CFC production by the year 2000. They added to the phaseout list methyl chloroform, carbon tetrachloride, and halons, which are also ozone-destroying chemicals. Several Third World countries refused to sign unless an international fund was established to help them with the technical shift to CFC alternatives. When the United States balked at contributing to that fund, the agreement almost failed, but in the end the fund was established. The reduction in stratospheric Cl now expected from the London Agreement is shown in Figure 5-6.

Getting Along without CFCs

While the diplomacy was going on, a burst of industrial creativity was coming up with hundreds of ways of reducing the release of existing CFCs and of finding substitutes for them.

Because of the 1978 ban in the United States, manufacturers had already discovered alternate aerosol propellants, most of which proved less expensive than CFCs. As atmospheric chemist Mario J. Molina said:

"In 1978, when the United States prohibited the use of CFCs as propellants in spray cans . . . experts said the ban would put a lot of people out of work. It didn't. In any case, the world cannot afford the consequences of continuing to release CFCs into the environment."[15]

Coolants in refrigerators and air conditioners used to be released to the air when those units were serviced or discarded. Now devices have been invented to recapture, purify, and reuse those coolants. Some alternative coolants to replace CFCs were already known (they had been used before CFCs were synthesized), and others are being developed.

Electronics and aeronautics firms have worked out substitute solvents for cleaning circuit boards and airplane parts, some of them involving simple water solutions. They have also reworked manufacturing processes to eliminate many washing steps entirely, with considerable economic savings. Several firms from the United States and Japan have formed a coalition to share their research on these adaptations with electronics manufacturers all over the world, without charge, in order to encourage the phaseout of CFC solvents.[16]

Chemical companies are coming up with hydrogenated CFCs (only 2% to 10% as destructive to the ozone layer) and other completely new compounds to substitute for specific uses of CFCs.

Insulating plastic foam is being blown with other gases; hamburgers are being wrapped in paper or cardboard; consumers are returning to washable ceramic coffee cups instead of throwaway plastic ones.

The world can get along without CFCs. Industry is adjusting to a complete phaseout of these important chemicals with much less expense and economic disruption than anyone would have guessed when the international negotiations began. Since CFCs are also greenhouse gases several thousand times as powerful as carbon dioxide, their phaseout will not only reduce ozone depletion; they will also help reduce the probability of global climate change.

Meanwhile news from the stratosphere keeps coming in. In the spring of 1991 NASA announced that new satellite measurements over the northern hemisphere showed ozone depletion occurring about twice as fast as expected. For the first time in 1991 depressed ozone levels over populated areas in North America, Europe, and Central Asia extended into the summer, when radiation damage is most likely to harm both people and crops. During the decade of the 1980s summertime ozone levels fell by 3% in the Northern Hemisphere and 5% in the Southern Hemisphere, 3 times faster than they had fallen during the 1970s.[17] And in the fall of 1991, the ozone hole over the South Pole was larger and deeper than ever before.

The Moral of the Story?

One can draw many possible lessons from the ozone story, depending upon one's temperament and political predilections. Here are the ones we draw:

- Political will can be summoned on an international scale to keep human activities within the limits of the earth.
- People and nations do not have to become perfect saints in order to forge effective international cooperation on difficult issues, nor is perfect knowledge or scientific proof necessary for action.
- A world government is not necessary to deal with global problems, but it is necessary to have global scientific cooperation, a global information system, and an international forum within which specific agreements can be worked out.
- Scientists, technologists, politicians, corporations, and consumers can react quickly when they see the need to do so—but not instantly.
- When knowledge is incomplete, environmental agreements need to be written flexibly and reviewed regularly. Constant monitoring is needed to report the actual state of the environment.
- All the major actors in the ozone agreement were necessary and will be necessary again: an international negotiator like UNEP; some national governments willing to take the political lead; flexible and responsible corporations; scientists who can and will communicate with policy makers; environmental activists to put on pressure; alert consumers willing to shift product choices on the

basis of environmental information; and technical experts to come up with adaptations that can make life possible, convenient, and profitable even when it is lived within limits.

Of course we can also see in the ozone story all the ingredients of the structure of an overshoot and collapse system—exponential growth, an erodable environmental limit, and long response delays both physical and political. It took thirteen years from the first scientific papers to the signing of the Montreal Protocol. It will take thirteen more years until the Montreal Protocol, strengthened in London, is fully implemented. It will take more than a century for the chlorine to be cleansed from the stratosphere.

This is a story of overshoot. Everyone hopes it will not be a story of collapse. Whether it will be or not depends on how erodable or self-repairable the ozone layer is, on whether future atmospheric surprises appear, and on whether humanity has acted, and will continue to act, in time.

chapter 6

Technology, Markets, and Overshoot

All the evidence suggests that we have consistently exaggerated the contributions of technological genius and underestimated the contributions of natural resources . . . . We need . . . something we lost in our haste to remake the world: a sense of limits, an awareness of the importance of earth's resources.

Stewart Udall[1]

The species *Homo sapiens* has been on Earth for one hundred thousand years. Human beings have organized themselves into civilizations for ten thousand years. They have experienced rapid population growth for at most three hundred years. During those last few centuries spectacular technical and institutional innovations—from the steam engine to democracy, from the computer to the corporation—have allowed the human economy to transcend apparent physical and managerial limits and keep on growing. Especially over the past few decades, the evolving industrial culture has implanted within the human mind the expectation of ever-continuing growth.

Therefore the idea that there might be limits to growth is for many people impossible to imagine. Limits are politically unmentionable and economically unthinkable. The society tends to assume away the possibility of limits by placing a profound faith in the powers of technology and the workings of a free market.

> By the successive substitution of technologies . . . real output can go on increasing without limit, without the cumulative consumption of any particular resource exceeding set limits.[2]

> We are confident that the nature of the physical world permits continued improvement in humankind's economic lot . . . indefinitely. Of course there are always newly arising local problems, shortages, and pollutions. . . . But the nature of the world's physical conditions and the resilience in a well-functioning economic and social system enable us to overcome such problems, and the solutions usually leave us better off than if the problem had never arisen; that is the great lesson to be learned from human history.[3]

The most common criticisms of the World3 model twenty years ago were that it underestimated the power of technology and that it did not represent adequately the adaptive resilience of the free market. We knew about technology and markets, of course. We assumed in World3 that markets function to allocate investment essentially with perfection. We built technical improvements into the model, such as birth control, resource substitution, the Green Revolution in agriculture. Possible future technical leaps we tested in model runs. What if materials are almost entirely recycled? What if land yield doubles again and yet again? What if pollution control could be made 4 or 10 times more effective?

Even with those assumptions, the model world overshoots its limits. Even with the most effective technologies and the greatest economic resilience we can believe possible, *if those are the only changes,* the model generates scenarios of collapse.

In this chapter we will explain why. Before we go on, however, we need to acknowledge that we are on dangerous terrain. We are talking about processes that are not only scientific subjects of study but also cultural articles of faith. Any qualification or doubt we express will be heard by some people as heresy. If we suggest that technology or markets have problems or limits, some will label us antitechnology or antimarket.

We are neither. We are technically trained, and we are technological enthusiasts.[4] We count on technical efficiencies to ease the human economy down below the planet's limits with grace and without sacrifice. We also respect the virtues of the market. Two of us have Ph.D.'s from a major business school; one of us has been president of another business school and is currently a business manager. We count on improvements in market signals, as well as in technology, to bring about a productive, prosperous, sustainable society. But we do not count on technology or markets by themselves, unchanged, unguided by any purposes or values beyond those that dominate the market, to bring a sustainable society into being.

Our qualified faith in technology and markets is based on our understanding of systems principles and systems behavior. It comes from the discipline of having to express in the world model exactly *what technology is,* and exactly *what markets do.* When one has to model these processes concretely, instead of making sweeping general claims for them, one discovers their functions and powers in the world system and also their limitations.

In this chapter we will:

- Describe technology and market feedback processes as we understand them and have modeled them in World3,
- Show computer runs in which we assume more and more effective technologies,
- Explain why instability, overshoot, and collapse are still the dominant modes of behavior in these runs, and
- End with a short case study about the world's fisheries, which demonstrates how technologies and markets in the present world are in fact contributing to the collapse of a valuable resource.

Technology and Markets in the "Real World"

What, "really," is technology? The ability to solve any problem? The source of all evil? The physical manifestation of the inventive genius of humankind? An increase in the amount that can be produced by an hour of labor or a unit of capital? The control of nature by humanity? The control by some people of other people with nature as their instrument?

Human mental models contain all these concepts of technology, and more.[5]

What, "really," is the market? Some would say it is simply the place where buyers and sellers come together to establish an exchange price that expresses the relative value of each commodity. Others would say the free market is a fiction invented by economists. Some people who have been deprived of a working market think of it as a magical institution that somehow delivers blue jeans and videocassette recorders in abundance. Or is the market the ability to own capital privately and to keep the returns? Or the most efficient means of allocating society's products? Or a device by which some people control other people with money as their instrument?

Out of this confusion of models, the idea that people most commonly have in mind when they say that technology and markets can forestall the limits to growth goes like this:

- A problem related to limits appears: a resource becomes scarce, or a dangerous pollutant begins to build up.
- The market causes the price of the scarce resource to rise relative to other resources, or the pollutant begins to exact costs that are reflected in the prices of products or services. (Here there is usually an admission that the market needs significant adjustment in order to capture pollution costs.)
- The rising price generates responses. It pays prospectors to go find more of the resource. It causes manufacturers to substitute a more abundant resource for the scarce one. It forces consumers to use fewer products containing the resource or to use the resource more efficiently. It induces engineers to develop pollution control devices, or to find places to sequester the pollutant away from human society, or to invent manufacturing processes that don't produce the pollutant in the first place.
- These responses on both the demand and the supply side compete in the market, where buyers and sellers collectively decide which technologies and consumption patterns solve the problem most quickly and efficiently at least cost. The society then adopts the best solutions and overcomes the scarcity, or reduces the damage from the pollutant.

Notice that this model does not rely solely on technology or solely on the market, but on a smooth interaction between the two. The market is needed to signal the problem, to direct resources toward its solution, and to select and reward the best solution. The technology is needed to solve the problem. The whole package has to be present. Without signals and guidance from the market, the technology will not be forthcoming. Without technical knowledge and ingenuity, the market's signals and guidance will produce no results.

Notice also that this model takes the form of a negative feedback loop—a chain of causation that acts to reverse a change, correct a problem, restore a balance. The resource scarcity is overcome. The pollution is cleaned up or sequestered. The society can continue to grow.

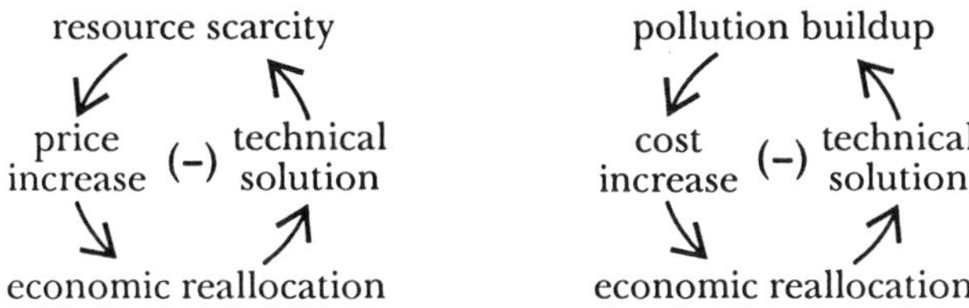

We believe that adjustment loops like these exist and are important. We have included them in many places in the World3 model, but not as a single, aggregate, wonder-working variable called "technology." Technologies arise from many causes and have many effects. Health care, for example, is automatic in World3. It is generated whenever the simulated world's service sector can pay for it. Birth control technology appears in World3 when the health care system can support it and when there is a demand for it in the form of a low desired family size. Development of land and improvements in land yield are also automatic in World3, as long as food demand is unsatisfied and capital is available.

If nonrenewable resources become scarce, the economy in World3 allocates more capital to discovering and exploiting them. We assume that the initial resource base can be exploited completely, though as resources are depleted it takes more and more capital to find and extract them. We also assume that nonrenewable resources are perfectly substitutable for each other, without cost or delay. (Therefore we lump them all together without distinguishing one from another.)

The market-technology adjustments mentioned above are built into World3. By changing numbers in the model we can strengthen or weaken them. If we don't change the numbers, these technologies evolve in the simulated world at roughly the same stages of industrial output per capita at which they appeared in the present highly industrialized countries.

In World3 the need for a built-in technology—health care, birth control, agricultural improvement, resource discovery and substitution—is signalled perfectly and without delay to the capital sector. Those technologies are implemented without delay as long as there is sufficient industrial or service output to make them possible. We do not represent prices explicitly, because we assume that prices are intermediary signals in an adjustment mechanism that works instantly and perfectly. We represent the mechanism without the intermediary. That assumption omits many delays and inaccuracies that occur in "real" market systems.

There are a number of other technologies in World3 that don't become effective unless we turn them on in test scenarios. They include resource efficiency and recycling, pollution control, and land erosion control. Twenty years ago we didn't consider these technologies as so established that they were technically proven and ready to ship to anyone in the world who could pay for them.[6] Therefore we programmed them so they could be activated at any future simulated time. For instance we might suppose the entire world makes a major commitment to recycling in 1995 or a concerted effort against pollution in 2005. These "turn-on" technologies require capital, and they come on only after a development and implementation delay, which is normally set at twenty years, unless we decide, as we will later in this chapter, to shorten it.

The reason to have a computer model is to try out different assumptions and to explore different futures. We can, for example, look at Scenario 2, the last run we showed in Chapter 4, where growth was ended by a pollution crisis, and we can ask: What if that simulated world responded to that rising curve of pollution by making a determined investment in pollution control technology?

Scenario 3 shows what happens.

Stretching the Limits with Technology in World3

In Scenario 3 and all further computer runs in this book, we continue to assume "double resources" (200 years worth at 1990 consumption rates) as in Scenario 2. That makes Scenario 2 the basis of comparison for the technology and policy changes that follow. We apply changes one at a time—first pollution control technology, then land yield technology, and so forth—not because we think there's any "realistic" possibility of the world trying out just one technology at a time, but because pedagogically that is the only way to make the model's responses understandable. In our own work with World3, even if we want to try three simultaneous changes, we apply them one at a time so we can understand the effect of each one separately before we try to comprehend the combined, and often interacting, effects of all at once.

In Scenario 3 we have assumed that in the simulated year 1995, long before the global pollution level has risen high enough to cause measurable damage on a global level to either health or crops, the world decides to bring pollution down to the levels that prevailed in 1975 and systematically allocates capital toward that goal. It chooses an "end of the pipeline" approach, abating pollution at the point of emission, rather than reducing throughput at the source.

We assume that it takes only twenty years for any new pollution abatement technology to be developed and installed worldwide. As the technology takes effect, it reduces the amount of pollution emitted with every unit of industrial output by up to 3% per year (depending upon need) until pollution is brought down to the relatively low level that prevailed in 1975. (That target was set arbitrarily; any target can be tried in model tests.)

In this scenario pollution continues to rise in spite of the abatement program, because of the delays in implementation, and because of continued underlying growth in agricultural and industrial production. But pollution stays much lower than it did in Scenario 2. It never gets high enough to affect human health significantly, but it does reduce land fertility after about 2015. Yields do not drop, because the reduction in the fertility of soils is compensated for by additional agricultural inputs. ("Real-world" examples of this phenomenon are the use of lime to compensate for acid rain, or the use of fertilizers to substitute for the lowered nutrient-generating capabilities of pesticide-poisoned soils.)

The countertrends of rising agricultural inputs and declining soil fertility in Scenario 3 lead to stagnant food production after the simulated year 2010. The population goes on growing, so food per capita begins to turn down. Total industrial output peaks and begins to decline by 2035, because so much capital has been pulled into the agricultural, resource, and pollution sectors that there is no longer enough investment to cover depreciation. Since population continues to grow until after 2050, per capita industrial output drops, the economy declines, and a collapse sets in.

The society in Scenario 3 greatly reduces its pollution levels but in the process suffers a food crisis. What if it also turns its technological powers to the problem of raising more food? One possible result is shown in Scenario 4.

In this model test the pollution abatement program of Scenario 3 is activated again and at the same time the world society decides in 1995, well in advance of a global food crisis, to increase agricultural yields even beyond the increases known technologies can provide. (The additional technologies could be, for example, genetic technologies, added to the continued spread of the chemical-based technologies of the twentieth century.) The new agricultural technologies are also assumed to take twenty years to implement worldwide, to cost something in terms of capital, and to be able to raise yields by up to 2% per year (again depending upon need). Note that annual gains of 2% per year achieved over a century would imply total increases in land yield by a factor of more than 7, if there were no technology development delays!

Scenario 3 Double Resources and Pollution Control Technology

In this scenario we assume doubled resources, as in Scenario 2, and also increasingly effective pollution control technology, which can reduce the amount of pollution generated per unit of industrial output by 3% per year. Pollution nevertheless rises high enough to produce a crisis in agriculture that draws capital from the economy into the agriculture sector and eventually stops industrial growth.

SCENARIO 3

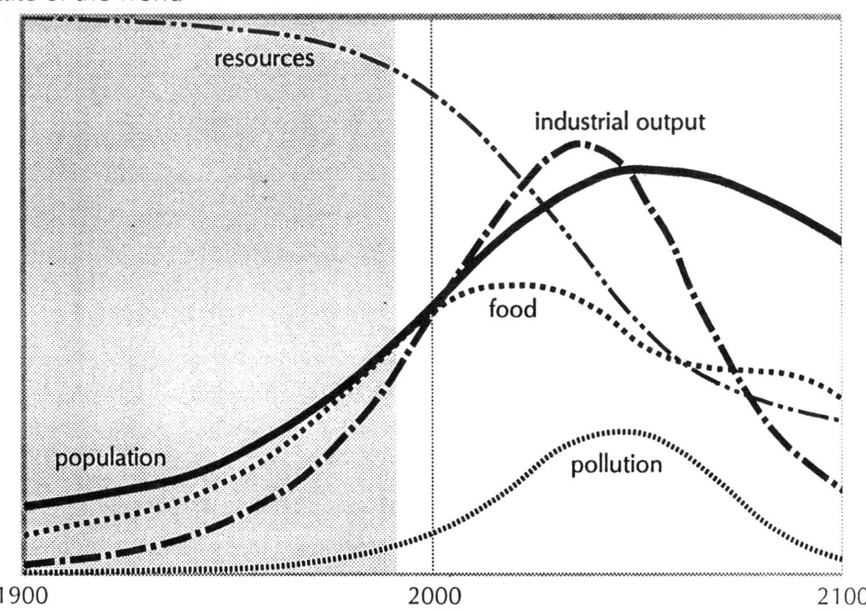
State of the world
resources
industrial output
food
population
pollution
1900
2000
2100

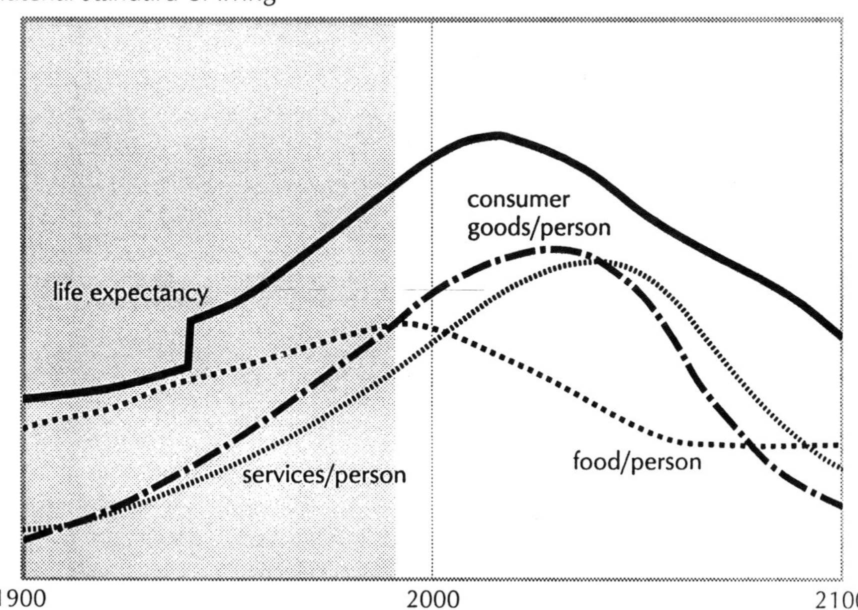
Material standard of living
consumer
goods/person
life expectancy
services/person
food/person
1900
2000
2100

This new agricultural technology, combined with the pollution reduction technology, in fact keeps land yield rising rapidly in Scenario 4, until average worldwide yields reach almost 4 times their 1990 level by 2100. Food production as a whole does not rise much higher or much longer than it did in the previous scenario, however, because the higher yields are being obtained from less and less land. The tremendously high agricultural intensity in this simulated world induces galloping land erosion. With less land, farmers work to get even higher yields from the land that is left, which causes still more erosion, and so forth, in a positive loop carrying the land system downhill rapidly. The stressed agricultural sector pulls more and more capital and human resources from the economy, at a time when the diminishing nonrenewable resource base is also demanding capital.

Surely, you would say, no sane society would pursue an agricultural technology that increases yields while destroying land. There are many examples of this behavior in the world today (for example the land lost to salt accumulation in the Central Valley of California while nearby land is pushed to ever-higher yields), but let us assume greater rationality on the part of coming generations. Let us add land protection technologies to the pollution control and yield-enhancing technologies.

No sooner said, in a computer world, than done. Scenario 5 shows the results of all those changes taking place at once. Here we assume, starting in 1995, the technical programs already described that reduce pollution output per unit of industrial production and raise land yield, and we add a program that reduces global land erosion by a factor of 3. The first two programs we assume require capital investment, the third we assume does not.

The result in Scenario 5 is a crisis not in resources, pollution, or land, but in all of them more or less at once. Food is sufficient, pollu-

Scenario 4 DOUBLE RESOURCES, POLLUTION CONTROL TECHNOLOGY, AND LAND YIELD ENHANCEMENT

If the model world adds to its pollution control technology a set of technologies to increase greatly the yield per unit of land, the high agricultural intensity speeds up land loss. The world's farmers are getting higher and higher yields on less and less land, and at an ever-higher cost to the capital sector.

SCENARIO 4

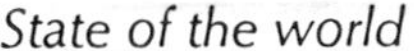

State of the world

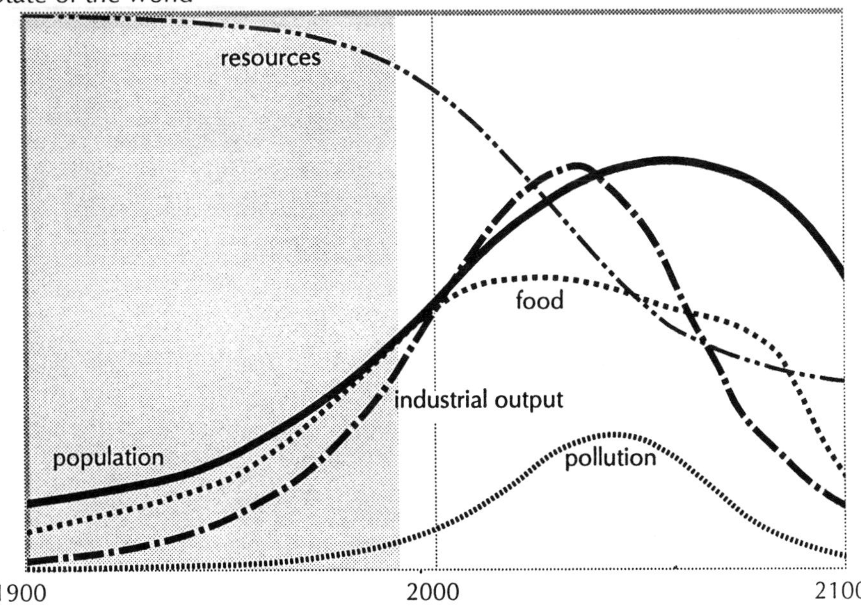

Material standard of living

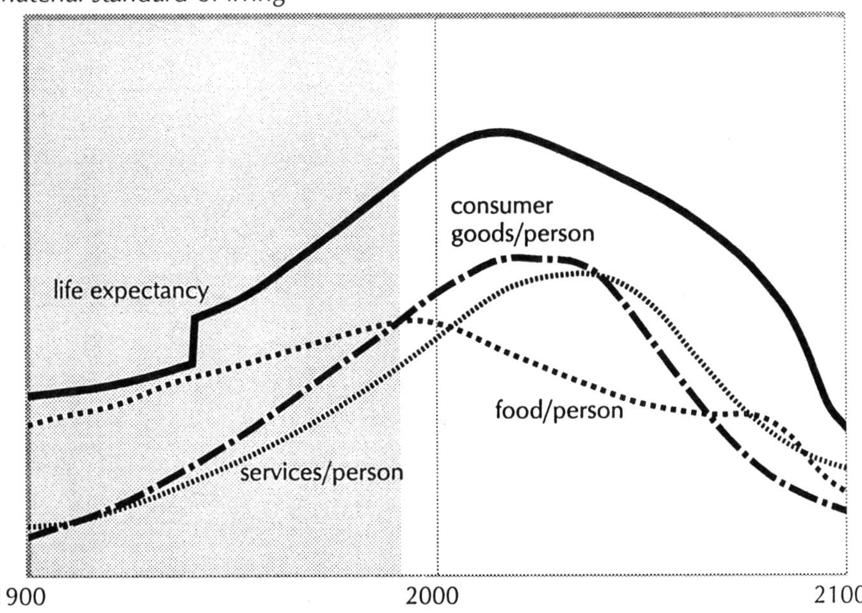

tion is tolerable, the economy grows, life expectancy rises until about 2020, when the costs of the various technologies, plus slowly rising costs of obtaining resources, simply demand more capital at any one time than the economy can provide.

One might argue about which priority a society stressed in so many directions would let drop first. Would it let land erode, reduce farm inputs, let pollution rise, or get along with reduced flows of raw materials? World3 assumes that materials would be given a high priority, in order to go on producing the industrial output needed to keep everything else going. That particular choice, and the exact model behavior after capital becomes insufficient is not important. We do not pretend to be able to predict what the world would do if it actually came to such a pass. The important point is simply that such a predicament is possible. It is one more way the overshoot and collapse behavior can manifest.

If resources are the final blow causing the collapse in Scenario 5, then a program of resource-saving technologies, added to all the others, should help. In Scenario 6 we start up in the simulated year 1995 a program to reduce the amount of nonrenewable resources needed per unit of industrial output by 3% per year until total resource consumption decreases to its approximate level in 1975 (this will also reduce pollution generation). We also still have pollution control technologies, land yield technologies, and land erosion control operating.

This combination of technologies permits the simulated world economy to go on growing smoothly through the middle of the twenty-first century. Nonrenewable resources are depleted only slowly; their cost remains low. Food production increases steadily. Pollution gets high enough to depress land fertility, but its effect can be overcome by additional agricultural inputs. Population levels off at about 10 billion.

Scenario 5 DOUBLE RESOURCES, POLLUTION CONTROL TECHNOLOGY, LAND YIELD ENHANCEMENT, AND LAND EROSION PROTECTION

Now a technology of land preservation is added to the agricultural yield-enhancing and pollution-reducing measures already tested. The result is further population and capital growth, which leads to a crisis not in resources, pollution, or land, but in all three at once.

SCENARIO 5

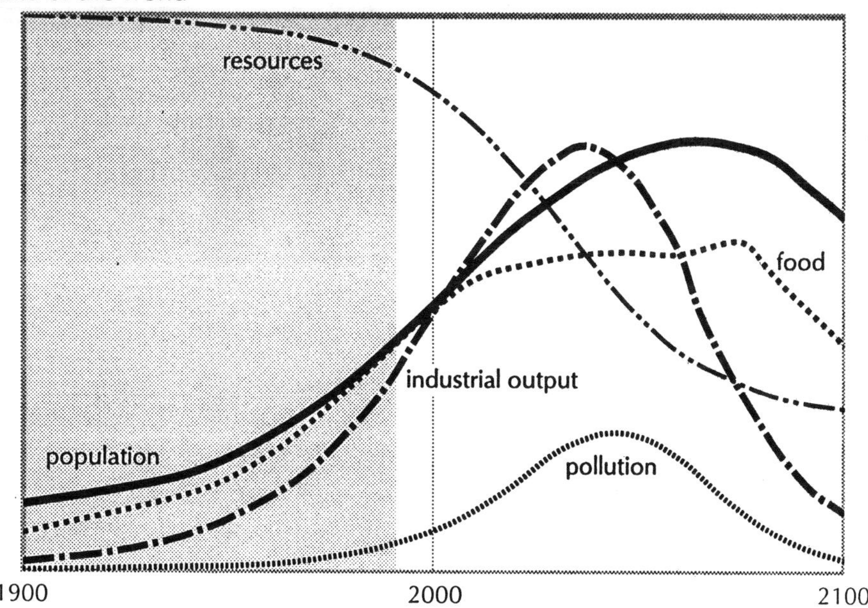
State of the world
resources
food
industrial output
population
pollution
1900
2000
2100

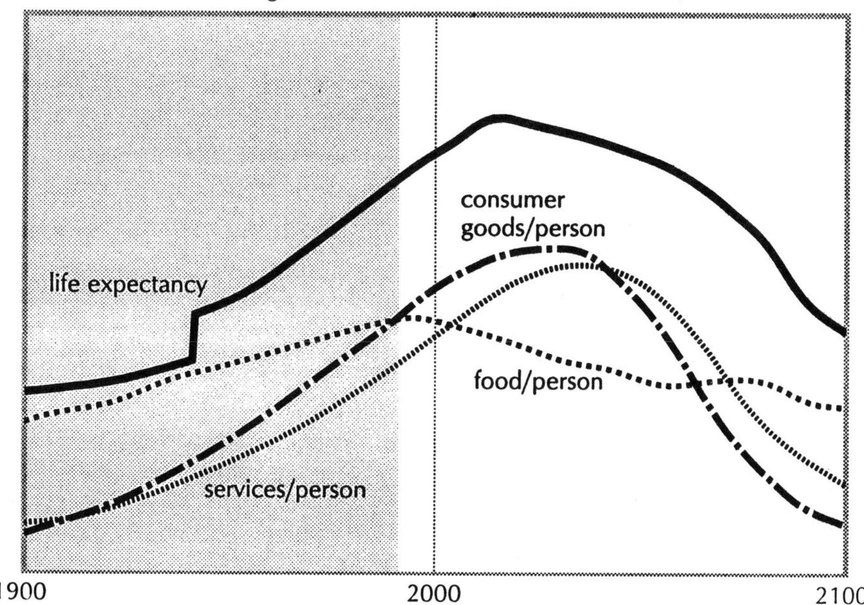
Material standard of living
consumer goods/person
life expectancy
food/person
services/person
1900
2000
2100

It does not level off because of a demographic transition in which birth rates come down to equal death rates. Rather, after 2020, death rates slowly rise to equal birth rates.

Why do death rates rise? Not because of any obvious crisis. Escalating technological effectiveness is successfully staving off sudden collapse. But the simulated world still overshoots its limits, as evidenced by the slowly declining industrial output and the steady erosion in material quality of life. Industrial output declines because the continued expense of protecting the population from starvation, pollution, erosion, and resource shortage cuts into the investment available for further growth.

After the simulated year 2020 in Scenario 6 life expectancy falls, first slowly and then rapidly, mainly because the declining economy cannot maintain high levels of health services. Food per capita stagnates; consumer goods per capita go down steadily after 2015. Because pollution works its way through the ecosystem only slowly and pollutants have long lifetimes, pollution goes on rising for two decades after its emissions have begun to decrease, though it never rises high enough to affect the life expectancy of the global population.

This is a society that is using its increasing technical capacity to maintain growth, while the growth eventually undermines the effects of those technologies. Ultimately the simulated world fails to sustain its living standards as its technology becomes too expensive and its environment degrades.

What if the technologies are brought on faster? What if the delays in their development and implementation are reduced from twenty

Scenario 6 DOUBLE RESOURCES, POLLUTION CONTROL TECHNOLOGY, LAND YIELD ENHANCEMENT, LAND EROSION PROTECTION, AND RESOURCE EFFICIENCY TECHNOLOGY

Now the simulated world is developing powerful technologies for pollution abatement, land yield enhancement, land protection, and conservation of nonrenewable resources all at once. All these technologies are assumed to cost capital and to take 20 years to be fully implemented. In combination they permit the simulated world to go on growing until 2050. What finally stops growth is the accumulated cost of the technologies.

Scenario 6

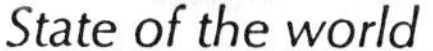

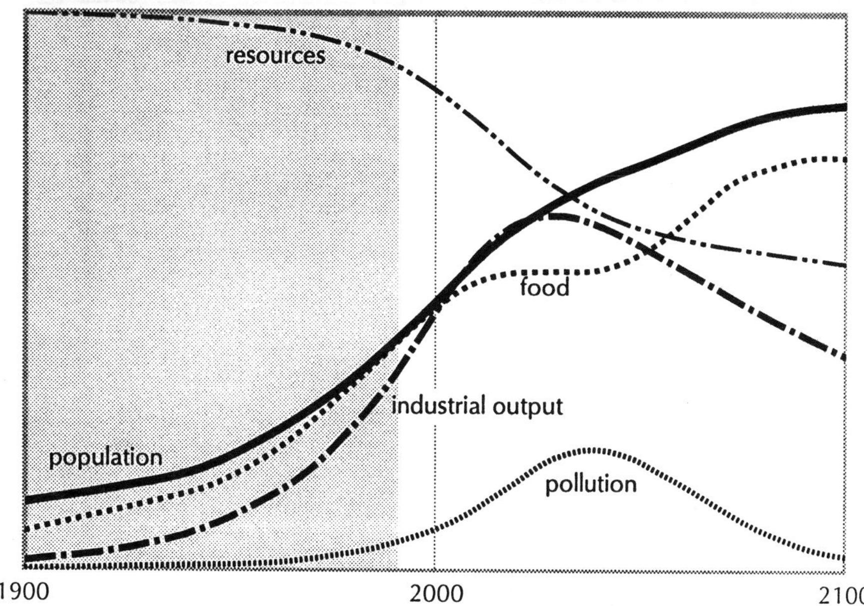

Material standard of living

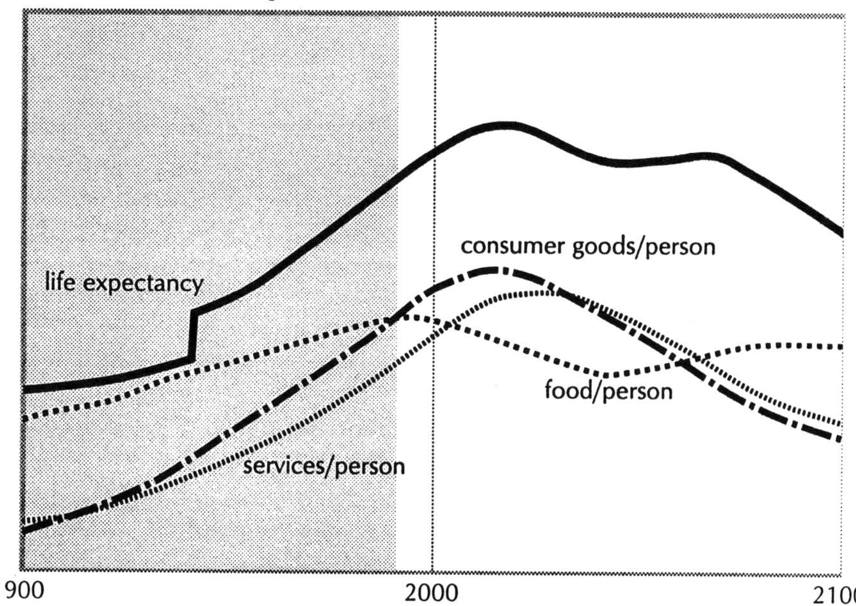

years to five years? Scenario 7 shows the results. Here we have pollution control, resource use reduction, land yield enhancement, and erosion control all beginning to be implemented worldwide by the year 1995 and increasing in effectiveness throughout the simulated twenty-first century. Each new device, each new process for reducing depletion or pollution is installed around the world within five years.

In this run industrial output grows for thirty years longer than it did in Scenario 6. The population grows to 12.5 billion while food per capita stays adequate, but not abundant. The average amount of food per capita remains slightly below 1990 world average levels; food production rises continuously but at just about the same rate as the population. Pollution stays quite low. Nonrenewable resources do not become scarce, though they are constantly decreasing. Consumer goods per person decline gradually after 2015; services per person decline after 2020; total industrial output stagnates after 2050 and declines after 2075.

The simulated world society in Scenario 7 is foresighted, highly technical, and frugal. By jumping out ahead of its problems before worldwide crises force it to, it manages to support a rising population at a decent standard of living throughout the twenty-first century. But in the second half of that century, its material quality of life gradually falls. Its large population living with moderate material throughputs is still stressing the limits of the earth. The increasing cost of holding off the limits stops and then depresses industrial growth.

After a session of working with a model, computer or mental, it's a good idea to step back for a moment and remember that it is not the "real world" you have been experiencing, but a representation of the world that is "realistic" in some respects, "unrealistic" in others. The

Scenario 7 All Technologies Applied with Shorter Delays

This model run is identical to the previous one, except that technology development is assumed to take only 5 years instead of 20 to have worldwide effect. Industrial output grows 20 years longer than it did in Scenario 6 and population becomes higher by 2 billion. But the material standard of living is falling slowly. The increasing cost of holding off the limits finally stops industrial growth.

SCENARIO 7

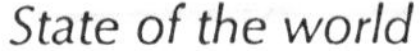

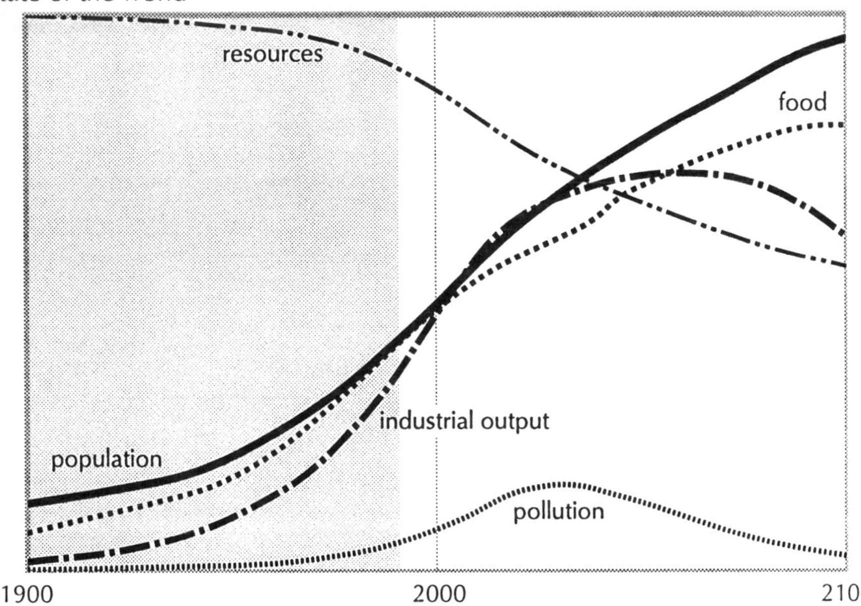

Material standard of living

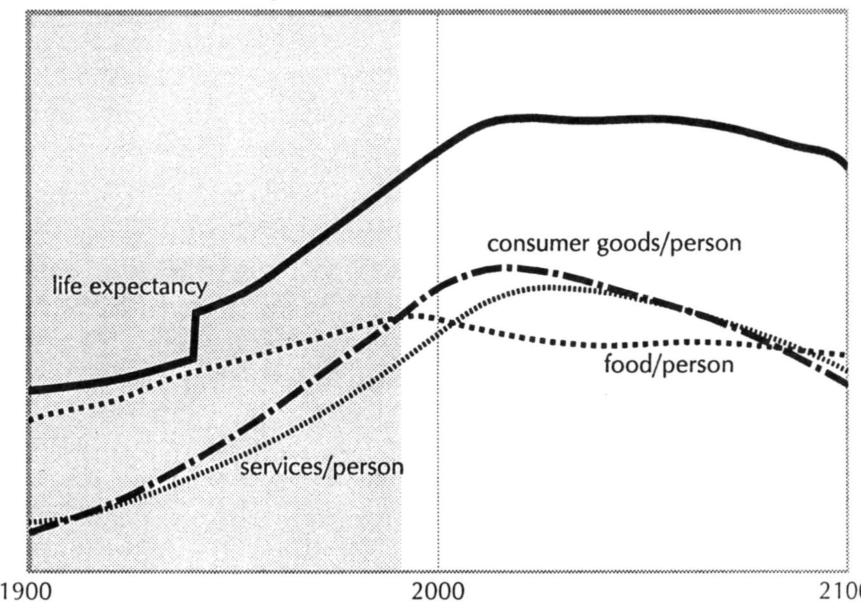

task is to find whatever insight is in the model and to assess where its "realism" ends and its uncertainties or deliberate simplifications begin. At the end of this series of computer runs, we need to stop and regain perspective.

World3, we must remember, does not distinguish the rich parts of the world from the poor. All signals of hunger, resource scarcity, and pollution are assumed to come to the world as a whole and to elicit responses that draw on the coping capabilities of the world as a whole. That may make the model too optimistic. In the "real world" if hunger is mainly in Africa, if pollution crises are mainly in Central Europe, if land degradation is mainly in the tropics, if the people who experience problems first are those with the least economic or technical capability to respond, that may mean very long delays before problems are corrected. Therefore the "real" system may not respond as forcefully or successfully as does the World3 system.

The model's perfectly working market and smooth, successful technologies (with no surprising "side effects") are also probably much too optimistic. So is the assumption that political decisions are made without cost and without delay.

We have to remember too that the World3 model has no military sector to drain capital and resources from the productive economy. It has no war to kill people, destroy capital, waste land, or generate pollution. It has no civil strife, no strikes, no corruptions, no floods, earthquakes, volcanic eruptions, Chernobyls, AIDS epidemics, or surprising environmental failures. In these senses it is wildly optimistic. The model could be representing the uppermost possibilities for the "real world."

On the other hand some people would say the technologies in the model are too limited. They would turn the technological cranks in the model much harder to bring the technologies on faster or even without limit (see Figure 4-7). Our assumptions about discoverable resources, developable land, and absorbable pollution may be too low. They may also be too high. We have tried to make them moderate, given the statistics available to us and our own assessment of technical possibilities.

With all these uncertainties, we obviously should not read out exact developments in the various scenarios with any quantitative precision.

We don't take it as significant, for instance, that a food crisis appears in Scenario 3 before a resource crisis. It could very well happen the other way around. We wouldn't swear that the world could actually support a population of 12 billion moderately well for fifty years, as it does in Scenario 7. We are not predicting an industrial turndown starting exactly in 2075. The numbers are just not good enough for World3 outputs to be read that way.

So what, if anything, can we learn from these technology modeling exercises?

Why Technology and Markets Alone Can't Avoid Overshoot

One lesson from these runs is that in a complex, finite world if you remove or raise one limit and go on growing, you encounter another limit. Especially if the growth is exponential, the next limit will show up surprisingly soon. There are *layers of limits.* World3 contains only a few. The "real world" contains many more. Most of them are distinct, specific, and locally variable. Only a few limits, such as the ozone layer or the greenhouse gases in the atmosphere, are truly global.

We would expect different parts of the "real world," if they keep on growing, to run into different limits in a different order at different times. But the experience of successive and multiple limits in any one place, we think, would unfold much the way it does in World3. And in an increasingly linked world economy, a society under stress anywhere sends out waves that are felt everywhere. Free trade enhances the likelihood that those parts of the world included in the free trade zone will reach limits simultaneously.

A second lesson is that the more successfully society puts off its limits through economic and technical adaptations, the more likely it is in the future to run into several of them at the same time. In most World3 runs, including many we have not shown here, the world system does not run out of land or food or resources or pollution absorption capability, it *runs out of the ability to cope.*

"The ability to cope" in World3 is represented, too simply, by a single variable: the amount of industrial output available each year to be invested in solving various problems. In the "real world" there are other components of the ability to cope: the number of trained people, the amount of political attention, the financial risk that can be handled, the institutional capacity, the managerial ability. All these capabilities can grow over time, if society invests in developing them. But at any one time, they are limited. They can process and handle just so much. When problems arise exponentially and in multiples, even though those problems could be dealt with one by one, the ability to cope can be overwhelmed.

Time is in fact the ultimate limit in the World3 model—and, we believe, in the "real world." The reason that growth, and especially exponential growth, is so insidious is that it shortens the time for effective action. It loads stress on a system faster and faster, until coping mechanisms that have been able to deal with slower rates of change finally begin to fail.

There are three other reasons why technology and market mechanisms that function well in a more slowly changing society cannot solve the problems generated by a society driving toward interconnected limits at an exponential rate. One is that these adjustment mechanisms themselves have costs. The second is that they themselves operate through feedback loops with information distortions and delays. The third is that the market and technology are merely tools that serve the goals, the ethics, and the time perspectives of the society as a whole. If the goals are growth-oriented, the ethics are unjust, and the time horizons are short, technology and markets can hasten a collapse instead of preventing it.

The *costs* of technology and the market are in resources, energy, money, labor, and capital. Those costs tend to rise nonlinearly as limits are approached, which is another source of surprises in systems behavior.

We have already shown in Figures 3-17 and 4-6 how the wastes produced and the energy necessary to extract nonrenewable resources rise spectacularly as the resource grade declines. Figure 6-1 shows two other typical rising cost curves: the cost per ton of abating the pollutants sulfur dioxide and nitrogen oxide as a function of the total amount removed from a smokestack or tail pipe. It is fairly inexpensive to remove

Figure 6-1 NONLINEAR COSTS OF POLLUTION ABATEMENT

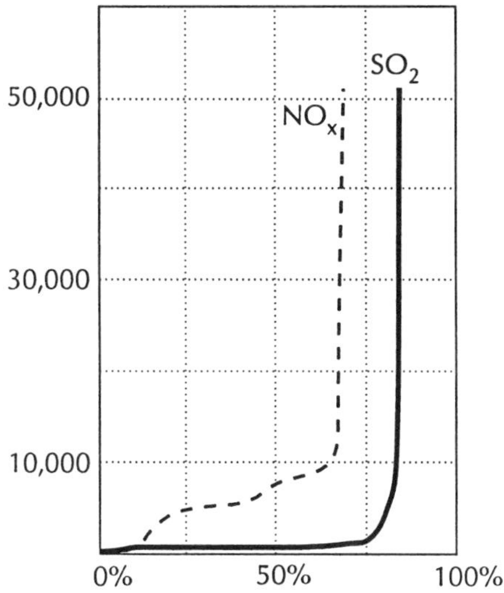

The air pollutants SO_2 and NO_x may be removed from smokestack gases to a significant degree at a low cost, but at some level of required abatement the cost of further removal rises precipitously. The cost curve for SO_2 removal here is calculated for Eastern Europe in deutsche marks (DM/ton); the NO_x curve is for Western Europe. (*Source*: *J. Alcamo et al.*)

almost 80% of the sulfur dioxide from smokestack emissions. There is a rising but still affordable cost for removing about 70% of the nitrogen oxides. But then there is a limit, a threshold, beyond which costs of further removals rise enormously.

It is possible that further technical developments will shift both these curves slightly to the right, making more complete cleanup affordable. But the curves will always have the same shape. There are fundamental physical reasons why abatement costs soar as 100% abatement is demanded. If the number of emission sources keeps growing, those rising costs will be encountered. It may be affordable to cut pollutants per car in half. But then if the number of cars doubles, it is necessary to cut

Figure 6-2 OPEC OIL PRODUCTION CAPACITY UTILIZATION AND WORLD OIL PRICE

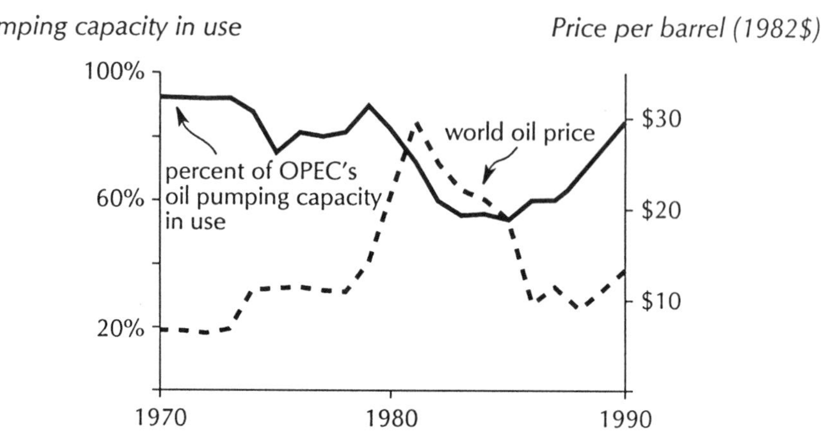

With most of OPEC's production capacity in use in the 1970s, small interruptions in oil supply precipitated sudden and extreme price changes. The oscillations in oil prices took more than 10 years to unfold and caused economic turbulence all over the world both on the way up and on the way down. (*Source*: *U.S. Department of Energy; Oil & Gas Journal.*)

pollutants per car in half again just to keep the same air quality. Two doublings will require 75% pollution abatement. Three doublings will require 87.5%, and by then the cost of further abatement is usually prohibitive.

That is why at some point it stops being true that growth will allow an economy to become rich enough to afford pollution abatement. In fact growth takes an economy along a nonlinear cost curve to the point where further abatement becomes unaffordable.

Delays in market and technology responses can be much longer than economic theories or other mental models lead one to expect. Technology-market feedback loops are themselves sources of overshoot, oscillation, and instability. One example of that, felt by all the world, was the oil price rise of the 1970s and early 1980s.

There were many causes of the "oil price shock" of 1973, but one of the most fundamental was the worldwide shortage of oil production capital (oil wells) relative to oil consumption capital (cars, furnaces, and other oil-burning machines). During the early 1970s the world's oil wells were working at over 90% capacity. Therefore a political upheaval in the Middle East that shut down even a small fraction of the world's oil production could not be met by increasing supply elsewhere. The market's only possible response was to raise the price.

That price rise, and a second one for the same reason in 1979 (see Figure 6-2), set off a wild set of economic and technical responses. On the supply side more wells were drilled and more pumping capacity was installed. Marginal oil deposits suddenly became profitable and were brought into production. The finding, building, and opening of oil production facilities, from wells to refineries to tankers, took time.

Meanwhile, consumers were reacting to the higher prices by conserving. Car companies came out with more efficient models. People insulated their houses. Electric companies shut down their oil-burning generators and invested in coal-burning or nuclear ones. Governments mandated various forms of energy saving and promoted the development of alternate energy sources. Those responses also took years. They ultimately resulted in long-lasting changes in physical capital.

It took nearly ten years before the many market responses, most of which had to be translated into physical production and physical capital, finally began to rebalance supply and demand. By that time conservation measures and new oil production capacity were coming on with such momentum that they overshot. By 1982 there was too much production capital compared to the decreasing needs of the consumption capital. OPEC began shutting down its pumping capacity. Its capacity utilization plunged from 90% to 50%. World oil price crept downward for four years and then in 1985 it plummeted.

Just as the price had gone too far up, now it went too far down. As oil production facilities shut down and oil-producing areas were struck with depressions, conservation efforts were abandoned. Designs for more efficient cars were put on the shelf. Investment in alternative energy sources dried up. After another ten years or so, as these adjustment mechanisms gather full steam going the opposite direction, they will set up the conditions for the next capital imbalance and the next oil price rise.

These overshoots and undershoots were a consequence of inevitable response delays in the oil market. They caused vast international shifts of wealth, enormous debts and surpluses, booms and busts and bank failures, all a result of trying to adjust the relative sizes of production capital and consumption capital for oil. None of these rises and falls in price were related to the actual underground quantity of oil (which was steadily going down) or to the environmental effects of drilling for, transporting, refining, and burning oil. The price signal provides information only about the relative scarcity or surplus of oil wells, not–until the very end of the depletion process–about the scarcity of oil.

Market signals such as oil price are too noisy, too delayed, too amplified by speculation, and too manipulated by private and public interest groups to give the world clear signals about oncoming physical limits. The market is blind to the long term and pays no attention to ultimate sources and sinks, until they are nearly exhausted, when it is too late to act. Economic signals and technological responses can evoke powerful responses, as the oil price example illustrates, but they simply are not connected to the earth system in the right places to give useful information about limits.

Finally there is the question of the *purposes* to which technology and markets are put. They are simply tools. They have no more inherent wisdom or farsightedness or moderation or compassion than does the human system that calls them forth. The results they produce in the world depend upon who wields them for what purpose. If they are used in the service of triviality, inequity, or violence, that is what they will produce. If they are asked to serve impossible goals, such as constant physical expansion on a finite planet, they will eventually fail. If they are called upon to serve feasible and sustainable goals, as we shall see in the next chapter, they can bring about a sustainable society.

Technological progress and market flexibility are essential tools for a sustainable society. When the world decided to get along without CFCs, technology made that change possible with amazing speed. When energy prices become undistorted by special interests and inclusive of environmental costs, the market will encourage the development of sustainable, affordable energy sources. We don't believe it is possible to bring about a sustainable world without technical creativity and entrepreneurship. But we don't believe they are sufficient. There are other human abilities that need to be called on to make the human world sustainable.

Technology, Markets, and the Destruction of Fisheries

"I remember catching 5000 pounds of fish in 8 nets. Today it might take up to 80 nets. Back then, the average codfish in the spring would probably be 25 to 40 pounds. Now it's 5 to 8 pounds." That comment by a fisherman on the Georges Bank fishing ground in the Northwest Atlantic[7] could be echoed, with different numbers and different fish species, by fishermen all over the world.

In 1990 the total world commercial marine fish catch declined by over 4 million tons. It was the first significant fall in world fish harvest since 1972. There is no way of knowing until many more years have passed whether this decline is a temporary hitch in a continuing growth curve, the first turndown in an overshoot-and-oscillation behavior, or the beginning of a collapse. But there is plenty of evidence of overfishing and even of fishery collapse at local scales. The U.N. Food and Agriculture Organization (FAO) believes the world's seas cannot sustain a commercial catch of more than 100 million metric tons per year from conventional resources—which is just about the level that the 1989 peak catch reached.

In nine of the nineteen world fishing zones monitored by the FAO, fish catches are above the lower limit of estimated sustainable yield.[8] In United States waters the National Fish and Wildlife Foundation says that 14 major fish species (which yield 20% of the world's fish harvest) are seriously depleted and would take 5 to 20 years to recover, even if all fishing stopped.[9] Populations of bluefin tuna, which normally live 30 years and grow to 1500 pounds, declined 94% in the 20 years between 1970 and 1990. The shrimp catch off the Florida Keys has declined from 6.4 million pounds per year to 2.4 million pounds.[10] Off the Kerala coast of India the fishing fleet is estimated to be 60% to 100% too high to be sustained.[11] The total catch from Norwegian waters is being sustained only by substituting less desirable commercial fish as the more desirable ones are being eliminated.

The General Fisheries Council for the Mediterranean says that:

> The influence of human activities on the marine environment . . . is suggested to be of particularly serious global concern at this time, with the most immediate effects . . . showing up first in enclosed or semi-enclosed bodies of water, such as the Mediterranean and the Black Sea. Here, growing populations and . . . industrial, agricultural and tourist activities are affecting . . . the fishery sectors, alongside the more classical fishery impacts, namely the largely uncontrolled fishing effort.[12]

The fishing industry around the world enjoys fairly free and vigorous markets, and it has seen in the past few decades extraordinary technological development. Refrigerated processing boats allow fleets to stay at distant fishing grounds without having to return home promptly with the harvest. Radar and sonar and satellite spotting bring boats to the fish with increasing efficiency. Driftnets 30 miles long allow economic large-scale fishing even in the deep seas. The result is that more and more fisheries are overshooting their sustainable limits. The technology being called forth is not that which enhances fish stocks, but that which seeks to find and catch every last fish.

> The trawler catch in New England peaked in 1983 and has since fallen sharply. Stocks of flounder and haddock are near record lows. The cod population is down. Bluefin tuna and swordfish have been depleted. Many New England fishermen find themselves in dire straits, victims of a get-it-while-you-can mentality. There have been booms and busts before. . . . But scientists say that this time is different, because the fleets are so big and the technology so good that fish no longer have anywhere to hide.[13]

Most people understand intuitively why that is happening. Fish are a common resource. The market gives no corrective feedback to keep competitors from overexploiting a commons; quite the contrary, it actively rewards those who get there first and take the most.[14] If the market signals scarcity by the rising price of fish, the richest people will still

Figure 6-3 BLUEFIN TUNA POPULATION DECLINES WHILE FISHING EFFORT CONTINUES TO RISE

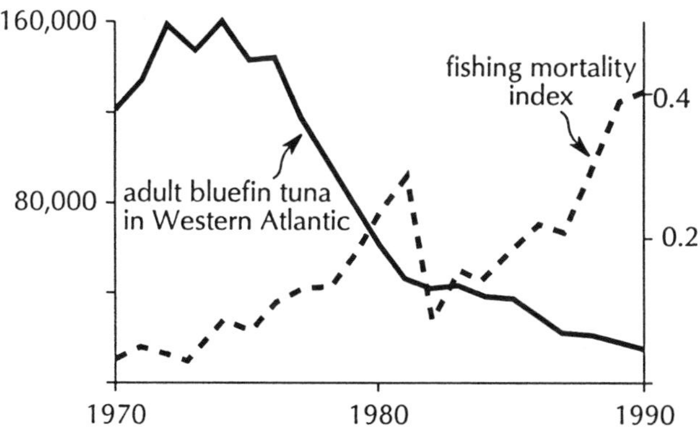

The Western Atlantic population of bluefin tuna over the age of 10 has been reduced by 94%. Because of the high value of these fish, the fishing effort continues. (*Source*: *International Commission for the Conservation of Atlantic Tuna.*)

be willing to pay that price. In Tokyo bluefin tuna may be worth as much as $100 a pound in the sushi market.[15] That high price does not signal scarcity and caution; it does not induce conservation; perversely it encourages more fishing effort as the bluefin population continues to be depleted (Figure 6-3). And the market does not allocate the fish to those who most need it for food, because the hungry have no power in the market.

> Increases in fish production have gone primarily to those countries that could afford to pay. . . . This trend is alarming because it indicates the potential danger that an increasing share of world catch will be siphoned off to the higher purchasing power in developed areas . . . leaving less fish where it is much needed–in the developing regions.[16]

Ecologist Paul Ehrlich once expressed surprise to a Japanese journalist that the Japanese whaling industry would exterminate the very source of its wealth. The journalist replied, "You are thinking of the whaling industry as an organization that is interested in maintaining whales; actually it is better viewed as a huge quantity of [financial] capital attempting to earn the highest possible return. If it can exterminate whales in ten years and make a 15% profit, but it could only make 10% with a sustainable harvest, then it will exterminate them in ten years. After that, the money will be moved to exterminating some other resource."[17] (A friend of ours has heard a similar argument from a firm cutting tropical timber in Sabah.)

The market players who are busily exterminating resources are utterly rational. What they are doing makes complete sense, given the rewards and constraints they see from the place they occupy in the system. The fault is not with people, it is with the system. An unregulated market system governing a common resource inevitably leads to overshoot and the destruction of the commons. Only political constraints of some kind can protect the resource, and those political constraints are not easy to attain.

> Fishing biologists have advised prohibiting all commercial cod catches in the Baltic next year because of decimated stocks. The Baltic Fisheries Commission, meeting in Warsaw, rejected this recommendation and attempted to agree on a reduced quota, but had to end their session with no agreement.[18]

It will take more than market and technical forces to save the world's fish resources from collapse. Used with no concept of limits, markets and technologies are instruments of overshoot. Used within limits and guided by long term communal values, however, the forces of the market and of technological development can help provide the world's fishing industry with rich harvests that can be sustained for generations.

A Summary

Exponential growth of population, capital, resource use, and pollution is still proceeding on the planet. It is propelled by attempts to solve keenly felt human problems, from unemployment and poverty to the need for status and power and self-acceptance.

Exponential growth can rapidly exceed any fixed limit. If one limit is pushed back, exponential growth will soon run into another.

Because of feedback delays, the global economic system is likely to overshoot and erode its sustainable limits. Indeed, for many sources and sinks important to the world economy overshoot has already occurred.

Technology and markets operate only with delays and only on imperfect information; they are themselves negative feedback processes with response delays that enhance the economy's tendency to overshoot.

Technology and markets serve the values of society or of the most powerful segments of society. If the primary goal is growth, they produce growth, as long as they can. If the primary goals are equity and sustainability, they can also serve those goals.

Once the population and economy have overshot the physical limits of the earth, there are only two ways back: involuntary collapse caused by escalating shortages and crises, or controlled reduction of throughput by deliberate social choice.

In the next chapter, we will see what happens when technological improvements are combined with deliberate social choices to limit growth.

chapter 7

Transitions to a Sustainable System

The stationary state would make fewer demands on our environmental resources, but much greater demands on our moral resources.

Herman Daly[1]

The human world can respond in three ways to signals that resource use and pollution emissions have grown beyond their sustainable limits.

One way is to disguise, deny, or confuse the signals: to build higher smokestacks, for instance, or to dump toxic chemicals secretly and illegally in someone else's territory; to overexploit fish or forest resources knowingly, claiming the need to save jobs or pay debts while in fact endangering the natural systems on which jobs and debt payments depend; to search for more resources while recklessly wasting those already discovered; to control prices that are rising in response to scarcity, or to put costs off onto the environment or onto faraway people or onto coming generations; to refuse to discuss population growth because the subject is too politically sensitive. These responses (and nonresponses) are refusals to deal with problems induced by limits, and they guarantee even worse problems in the future.

A second way to respond is to alleviate the pressures from limits by technical or economic fixes without changing their underlying causes: to reduce the amount of pollution generated per mile of driving or per kilowatt of electricity generation; to search for more resources, use resources more efficiently, recycle resources, or substitute one resource for another; to replace functions that nature used to perform, such as sewage treatment or flood control or soil fertilization, with human capital and labor; to develop better birth control pills. These measures are urgently needed. Most of them will ease some pressures for a while. But they do nothing about the underlying causes of the pressures.

The third way to respond is to step back and acknowledge that the human socioeconomic system as currently structured is unmanageable, has overshot its limits, and is headed for collapse, and, therefore, to *change the structure of the system.*

In everyday language, the phrase "changing structure" has a number of imprecise connotations, most of them ominous. It is used by revolutionaries to mean throwing people out of power, usually exacting bloody retribution in the process. Some people think of changing structure as changing *physical* structure, tearing down and rebuilding, constructing a new world. Most people fear that changing structure will be difficult, expensive and threatening to their security.

In systems language "changing structure" has a precise meaning that has nothing to do with throwing people out, tearing things down, or spending money. In fact doing any of those things without *real* changes in structure clearly will just result in different people spending as much or more money in a new system that produces the same old results.

In systems terms changing structure means changing the *information* links in a system: the content and timeliness of the data that actors in the system have to work with, and the goals, incentives, costs,and feedbacks that motivate or constrain behavior. The same combination of people, institutions, and physical structures can behave completely differently, if its actors can see a good reason for doing so and if they have the freedom to change. In time a system with a new information structure can socially and physically transform itself. It can develop new institutions, new rules, new buildings, people trained for new functions. That transformation can be natural, evolutionary, and peaceful.

Pervasive changes unfold spontaneously from new information structures. No one need engage in sacrifice or in strong-arming, except, perhaps, to get some people to stop deliberately confusing or distorting or ignoring information. Human history is full of structural transformations, the two most profound of which have been the Agricultural Revolution and the Industrial Revolution. In fact it is the success of those past transformations that has brought the world to the necessity for the next one.

World3 cannot begin to represent the evolutionary dynamics of a world system that is structuring itself in a new way. But it can be used to test some of the simplest changes that might result from a society that decides to restructure itself to reduce the probability of overshoot and collapse.

In the previous chapter we used the World3 model to see what happens if the world makes *quantitative,* not *structural,* changes. We put into the model higher limits, shorter delays, faster and more powerful technical responses to the delays, and weaker erosion loops. If we had taken those structural features away entirely—no limits, no delays, no erosion—we would have eliminated the overshoot and collapse behavior entirely (as we did in Figure 4-7, the "Infinity In, Infinity Out" run). But limits, delays, and erosion are physical properties of the planet. Human beings can mitigate them or enhance them, manipulate them with technologies, and live within them with many numerical degrees of freedom, but human beings cannot make them go away entirely.

The structural causes of overshoot over which people have the most power, we believe, are the only ones we did not change in Chapter 6, namely those that cause exponential growth in the human population and economic system. They are the social norms, goals, incentives, and costs that cause people to want more than a replacement number of children. They are the cultural expectations and practices that maldistribute income and wealth, that make people see themselves primarily as consumers and producers, that associate social status with material accumulation, and that define human goals in terms of getting *more* rather than having *enough.*

In this chapter we will change the driving positive loops that encourage exponential growth in the world system. We will come at the question of how to ease down from the state of overshoot from a new direction, beginning not with limits, delays, or erosion, but with the structural forces that cause growth.

Deliberate Constraints on Growth

Suppose that, starting in 1995, all couples in the world understood the implications of further population growth for the welfare of their own children. Suppose all people were assured by their societies of acceptance, of respect, of material security, and of care in their old age, no matter how many children they had. Suppose further that it became a social norm to raise every child with the highest possible standards of nutrition, shelter, health care, and education. Suppose as a consequence of these suppositions that all couples decided to limit their family size to two surviving children (on average), and that they had readily available fertility control technologies that allowed them to achieve that desired family size.

These changes would entail shifts in perceived costs and benefits, an increase in time horizon, an ability to see the social whole, an availability of new powers and choices and responsibilities–in short an informational restructuring equivalent to the one that has already brought down the birth rates of many populations in the industrial world.

If just that change is made in World3 and no others, the results are shown in Scenario 8, which should be compared with Scenario 2.

To generate this scenario we have set the average desired family size of the model population at two children and the birth control effectiveness at 100% after the simulated year 1995. As a result the model world's population growth moderates greatly, but age structure momentum carries the population to 6 billion in the simulated year 2000 and 7.4 billion in 2040. Because of the slower population growth rate, consumer goods per capita, food per capita, and life expectancy rise higher and stay high longer than they did in Scenario 2. Less industrial output is needed for the consumption and service needs of a growing population, so more investment is available to the capital sector. Therefore the simulated world's total industrial output grows faster and higher than it did in Scenario 2.

For a while average industrial output per capita in this run exceeds 1968-$500 per person per year—about the 1990 world average—but industrial output peaks and collapses at roughly the same time as it did in Scenario 2, and for the same reasons. The larger industrial plant emits more pollution and uses more resources. Pollution reduces agricultural yields. Capital has to be diverted to agriculture to keep food production going. After the year 2030 more capital also must be diverted to the nonrenewable resource sector to find and process scarce and depleted resource deposits.

Given the limits and technologies assumed in the simulated world of Scenario 8, that world cannot sustain 7.4 billion people with an ever-increasing per capita industrial output.

So what if the world's people decide to moderate not only their demand for children, but also their economic demands? What if they set themselves a goal of a simple but adequate material standard of living and, when they achieve that goal, they turn their attention to other, nonmaterial, nonconsuming pursuits? This, too, is a hypothetical information change, a change not in the physical world, but in peoples' heads (an enormous one, we realize). It means that people define their purposes, establish their status, challenge themselves with goals other than ever-increasing production and ever-accumulating material wealth.

Scenario 9 shows a simulated world again with a desired family size of two children and perfect birth control, and also with a definition of "enough." This world has decided to aim for an average consumer goods per capita of 1968-$350 per person per year—about the equivalent of that in South Korea, or about twice the level of Brazil in 1990.

Scenario 8 WORLD ADOPTS STABLE POPULATION GOALS IN 1995

This scenario supposes that after 1995 all couples decide to limit their family size to two children and have access to effective birth control technologies. Because of age structure momentum, the population continues growing well into the 21st century. The slower population growth permits industrial output to rise faster, until it is stopped by depleting resources and rising pollution.

SCENARIO 8

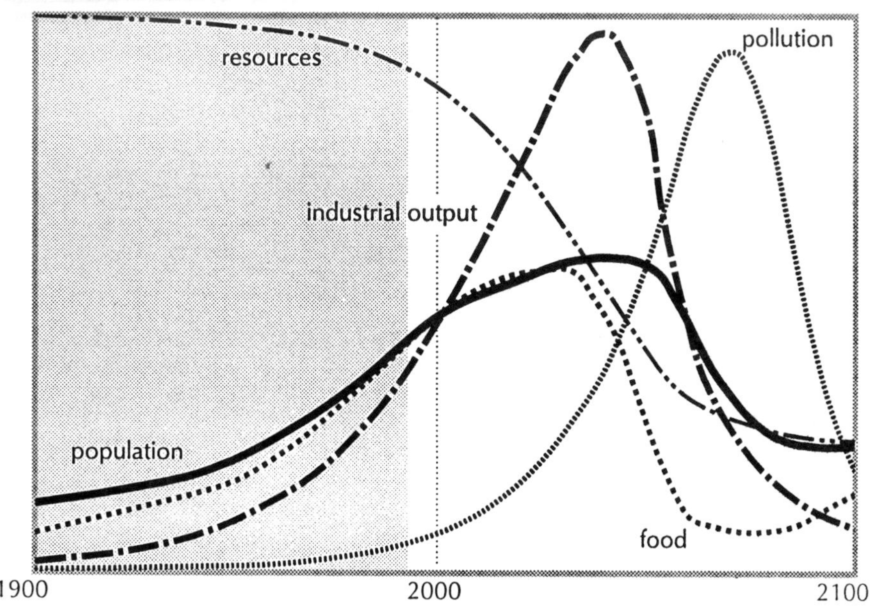
State of the world
resources
pollution
industrial output
population
food
1900
2000
2100

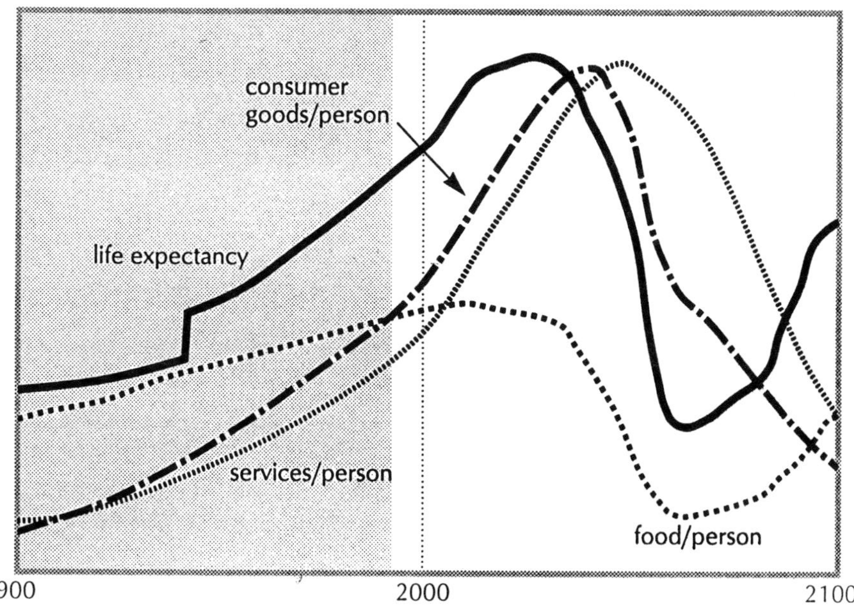
Material standard of living
consumer goods/person
life expectancy
services/person
food/person
1900
2000
2100

These comparisons should not be taken too literally, because a stabilizing society like the one in the model would be different in many ways from a growing society with an equivalent GNP today. At any industrial level it would have a higher fraction of consumption goods because it would have to allocate less economic output to investment for growth and less to defending against or compensating for resource depletion and pollution. If this hypothetical society could also reduce military expenditures and corruption, a stabilized economy with a consumer goods per capita of 1968-$350 could perhaps be equivalent in material comforts to the average level in Europe in 1990.

When the desired level of per capita production of the population in Scenario 9 is reached, investment in the capital sector is no longer needed for growth, only for offsetting depreciation. Depreciation is lower, because the average lifetime of capital is being increased by one-fourth in this scenario. The investment thus freed up is allocated to services, food, or resources, as needed.

The world of Scenario 9 manages to support its 7.3 billion people at its desired standard of living for almost fifty years, from 2005 to 2050. Consumer goods per capita rises 70% higher than its 1990 value; this could be a world with excellent education and health care for everyone. Total food production reaches a peak shortly after the year 2010, however, and falls steadily thereafter because of pollution, which rises steadily until the year 2075. More and more ameliorative investments in the agriculture sector are necessary. For a while they are also available. Finally, however, after about 2035, the depleting nonrenewable resource sector also begins to demand more capital, and the industrial sector cannot maintain itself.

Scenario 9 WORLD ADOPTS STABLE POPULATION AND INDUSTRIAL OUTPUT GOALS IN 1995

If the population adopts both a desired family size of two children and a deliberately moderated goal for industrial output per capita, it can maintain itself at a material standard of living 50% higher than the 1990 world average for almost 50 years. Pollution continues to rise, however, stressing agricultural land. Per capita food production declines, eventually carrying down life expectancy and population.

SCENARIO 9

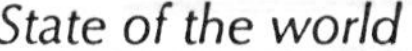

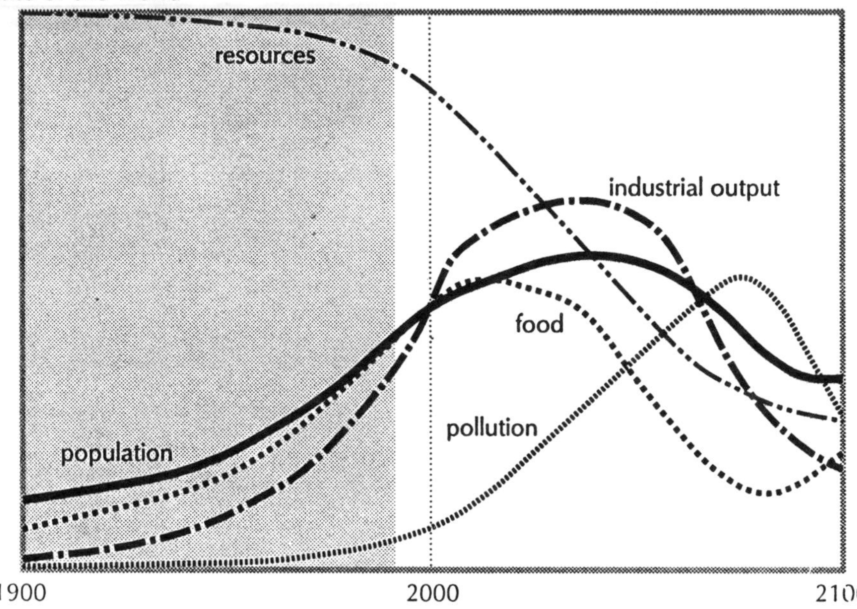

Material standard of living

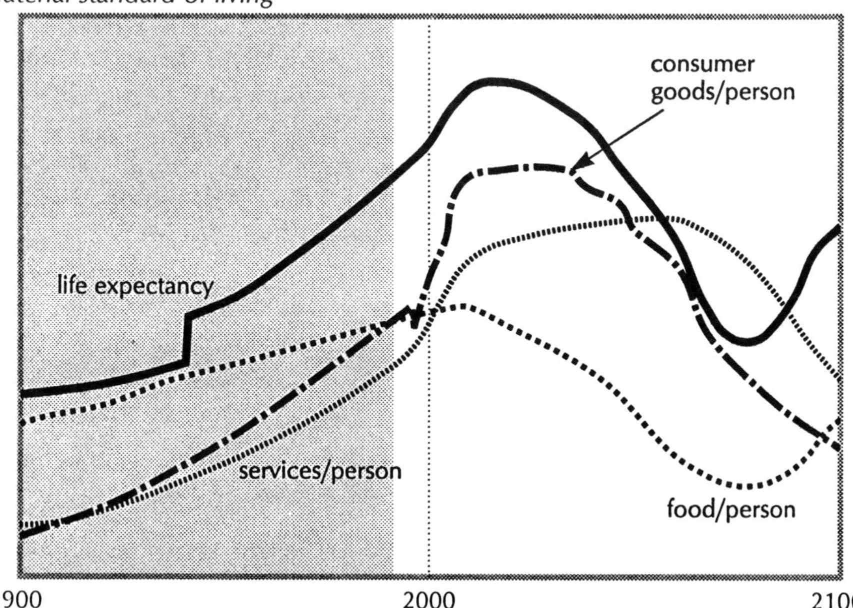

The simulated society in this scenario manages to sustain its desired standard of living with regard to material goods for two generations, but during that time its environment and food supply are steadily deteriorating. It needs to augment its social policies with more technological power.

Constraints on Growth plus Improved Technologies

In Scenario 10 the model world again decides on an average family size of two children starting in 1995, has perfect birth control effectiveness, and aims for consumer goods per capita of 1968-$350, all as in the previous model run. Furthermore, starting in 1995, it begins to employ the same technologies we tested in Chapter 6. These technologies increase the efficiency of resource use, decrease pollution emissions per unit of industrial output, control land erosion, and increase land yields until food per capita reaches its desired level.

We assume in Scenario 10, as we did in Chapter 6, that these technologies come on only when needed and only after a development delay of twenty years, and that they have a capital cost. In the simulations of Chapter 6, there wasn't enough capital to keep the technologies going while dealing with the various crises the rapidly growing society was running into. In the more restrained society of Scenario 10, where capital does not have to go either toward further growth or toward ameliorating a spiraling set of interacting problems, the new technologies can be fully supported. Operating steadily over a century, they reduce nonrenewable resource use per unit of industrial output by 80% and pollution production per unit of output by 90%. The slow growth in land yield pauses slightly in the early twenty-first century as pollution

Scenario 10 STABILIZED POPULATION AND INDUSTRY WITH TECHNOLOGIES TO REDUCE EMISSIONS, EROSION, AND RESOURCE USE ADOPTED IN 1995

In this scenario population and industrial output per person are moderated as in the previous model run, and in addition technologies are developed to conserve resources, protect agricultural land, increase land yield, and abate pollution. The resulting society sustains 7.7 billion people at a comfortable standard of living with high life expectancy and declining pollution until at least the year 2100.

SCENARIO 10

State of the world

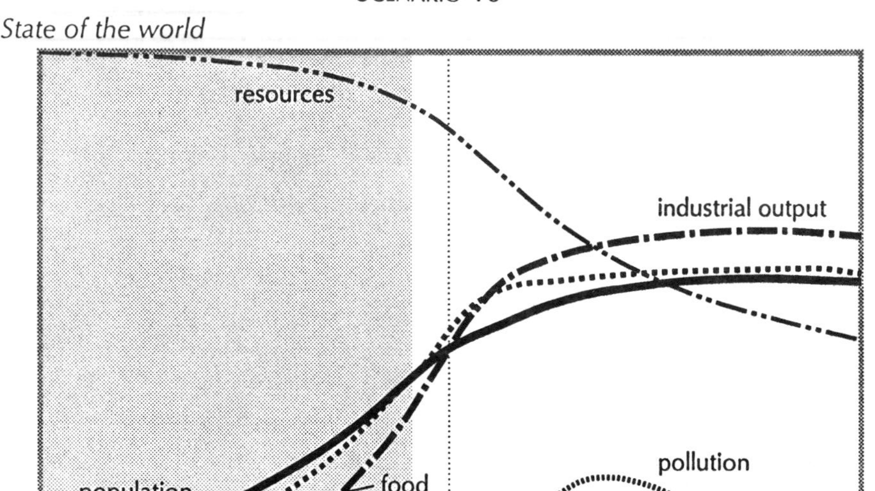

Material standard of living

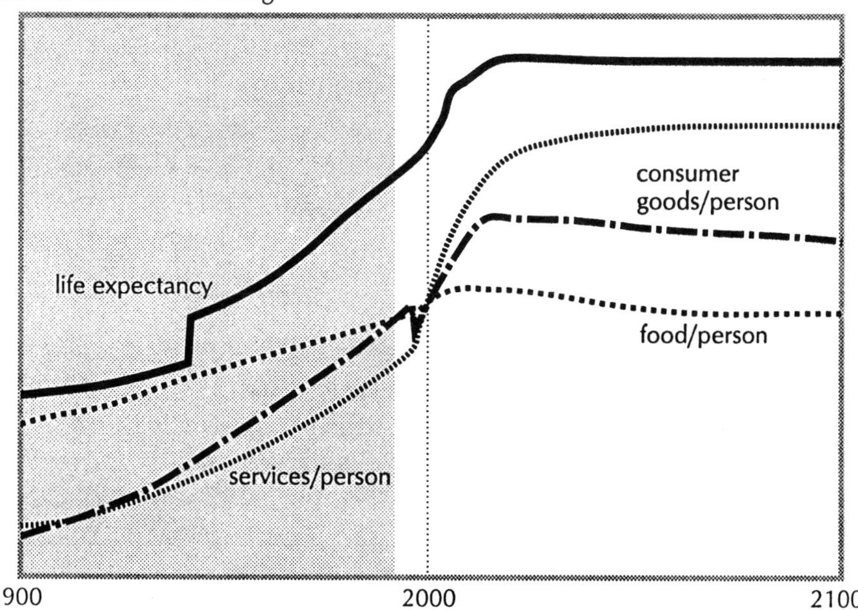

rises (a delayed effect of the pollution emissions around the end of the twentieth century), but by 2040 pollution begins to go down again. Land yield recovers and rises slowly for the rest of the century.

In Scenario 10 the population levels off at just under 8 billion and lives at its desired material standard of living for almost a century. After 2010 its average life expectancy stays at just over eighty years, its services per capita rise 210% above their 1990 level, and there is sufficient food for everyone. Pollution peaks and falls before it causes irreversible damage. Nonrenewable resources deplete so slowly that half the original endowment is still present in the simulated year 2100.

The society of Scenario 10 manages to begin reducing its total burden on the environment shortly after the year 2040. The rate of extraction of nonrenewable resources falls after 2010. Land erosion is reduced abruptly after 2040. The generation of persistent pollutants peaks in about 2015. The system brings itself down below its limits, avoids an uncontrolled collapse, maintains its standard of living, and holds itself nearly, but not quite, in equilibrium.

The word *equilibrium* in systems language means that positive and negative loops are in balance and that the system's major stocks–in this case population, capital, land, land fertility, nonrenewable resources, and pollution–are held fairly steady. It does not necessarily mean that the population and economy are static or stagnant. They stay roughly constant the way a river stays roughly constant, though new water is always running through it. In an equilibrium society like the one in Scenario 10, people are being born while others are dying; new factories, roads, buildings, machines are being built while old ones are being demolished (and recycled). Technologies are improving, and the steady flow of material output per person would almost certainly be changing and diversifying in content.

As a river may have ups and downs around some average flow, so could an equilibrium society vary, either by deliberate human choice or by unforeseen opportunities or disasters. As a river, when its pollution load is diminished, can purify itself and support more rich and varied aquatic communities, so can a sustainable society purify itself of pollution, acquire new knowledge, make its production processes more efficient, shift technologies, improve its own management, make distribution more equitable, and diversify itself. We think it is likely to do all those things, when the strains of growth are alleviated.

The sustainable society shown in Scenario 10 is one that we believe the world could actually attain, given the knowledge about planetary systems available to us. It has 7.7 billion people, and enough food, consumer goods, and services to support every one of them in material comfort. It is expending considerable effort and employing continually improving technology to protect its land, reduce its pollution, and use its nonrenewable resources with high efficiency. Because its growth slows and eventually stops, its problems are manageable and are being managed.

We think that is a picture not only of a feasible world, but of a desirable one, certainly more desirable than the simulated worlds of the previous chapter, which keep on growing until they are stopped by multiple crises. Scenario 10 is not the only sustainable outcome the World3 model can produce. Within the system's limits there are tradeoffs and choices. There could be more food and less industrial output or vice versa, more people living with a smaller stream of industrial goods or fewer people living with more. The world society could take more time to make the same transition to equilibrium than we have supposed in Scenario 10—but that would have its costs.

Delaying the transition to a sustainable world has already had its costs.

The Difference Twenty Years Can Make

In the next two runs we ask: What if the model world had undertaken the sustainability policies shown in Scenario 10 (desired family size of two children, desired consumer goods per capita of 1968-$350, all technologies of resource efficiency and pollution control) not in 1995 but in 1975? And what if it undertakes them not in 1995 but in the year 2015? What difference does plus or minus twenty years make?

Scenario 11, is exactly equivalent to Scenario 10 except that the policy changes are applied not in 1995 but in 1975. The differences between this world and the one in the previous scenario are subtle. Moving to sustainability twenty years sooner has produced a more secure and wealthy world, but it is not a qualitatively different one. The population levels off at 5.7 billion instead of almost 8 billion. Pollution peaks at a lower level and fifteen years sooner, and it interferes with crop yields much less than it did in Scenario 10. Life expectancy surpasses eighty years and stays high. There are more nonrenewable resources left by the end of the twenty-first century, and it takes less effort to find and extract them.

The Scenario 11 population reaches its desired level of industrial output per person sooner and is able to maintain it and support its improving technologies with no problems. This society has a more pleasant environment, more resources, more degrees of freedom; it is further from its limits, less on the edge than the society in Scenario 10.

That was twenty years backward, a future that might have been available once but is no more.

Twenty years forward makes a bigger difference, as one might expect, knowing the mathematics of exponential growth. In Scenario 12 we implement in World3 the same sustainability policies not in 1975, or 1995, but in 2015. By that time it is too late to avoid some heavy turbulence.

The simulated population in Scenario 12 reaches 8.7 billion. Though the same standard of living is set as a goal, to provide it to nearly a billion more people, industrial production must rise higher than it did in Scenario 10. The added industrial activity, plus the twenty-year delay in implementing pollution control technologies, brings about a pollution crisis, even though pollution control technologies are evolving in the same way they did in Scenario 11. Pollution re-

Scenario 11 STABILIZED POPULATION AND INDUSTRY WITH TECHNOLOGIES TO REDUCE EMISSIONS, EROSION, AND RESOURCE USE ADOPTED IN 1975

This simulation includes all the changes that were incorporated into the previous one, but the sustainability policies are implemented in the year 1975 instead of 1995. Moving toward sustainability 20 years sooner would have meant a considerably lower final population, less pollution, more nonrenewable resources, and a slightly higher material standard of living.

SCENARIO 11

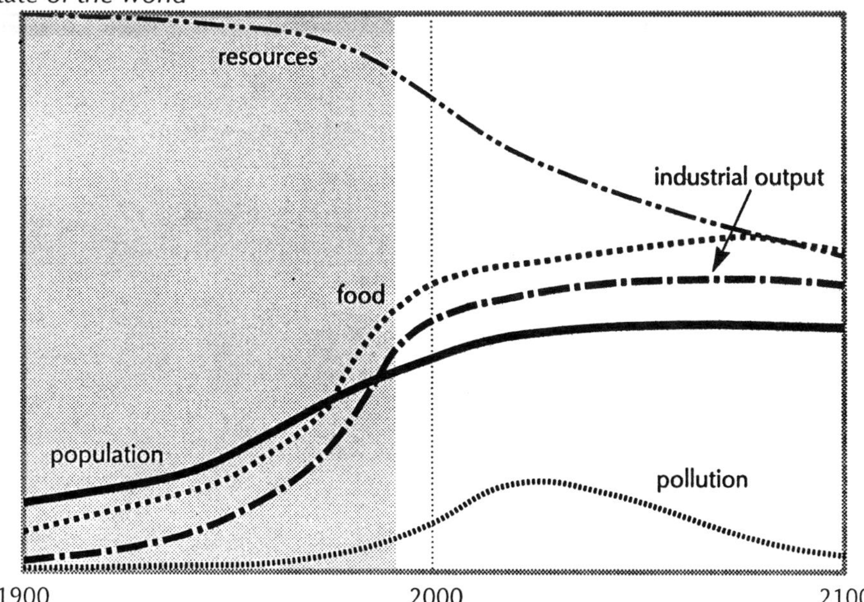

Material standard of living

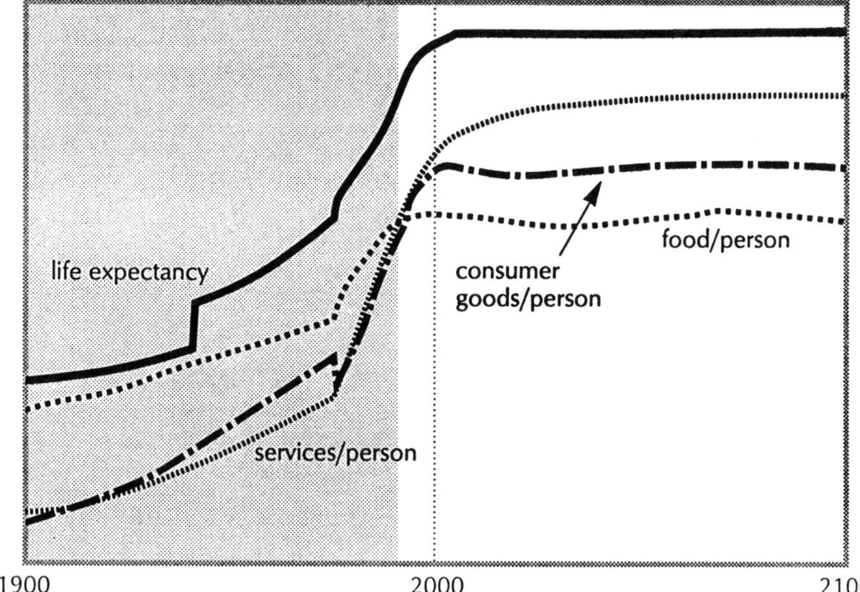

duces land yield, food per capita falls, life expectancy falls, and the population falls as well, to 7.4 billion.

With the smaller population and the ever-improving technologies (it may not be a "realistic" assumption that they would go on improving, given the economic decline in this model world), the world of Scenario 12 does eventually recover. After 2055 pollution begins to go down, food production revives, life expectancy rises once again. So much capital investment has been needed to correct the overshoot, however, that not enough is available to sustain the desired material standard of living. Consumer goods per capita peaks at 1968-$350 in about the year 2025 and falls slowly thereafter to half that level. The twenty-year delay in moving toward sustainability has greatly reduced the equilibrium standard of living that this simulated world can maintain.

How High Is Too High?

Scenario 12 shows what happens if the model society waits to make the transition to sustainability. Scenario 13 shows what happens if it aims just a bit too high.

Scenario 13 is directly comparable to Scenario 10 in that this world also begins to moderate its population and economy in 1995 and develops the same resource-conserving and pollution-reducing technologies. This time, however, the model world's goal for food per capita is set 50% higher and the goal for consumer goods per capita is set at 1968-$700 rather than 1968-$350, about 3.5 times the 1990 world average. It turns out that this combination of goals cannot be sustained for the population of nearly 8 billion people.

Consumer goods per capita never reaches its goal. It peaks below

Scenario 12 STABILIZED POPULATION AND INDUSTRY WITH TECHNOLOGIES TO REDUCE EMISSIONS, EROSION, AND RESOURCE USE ADOPTED IN 2015

Waiting to implement the sustainability policies until the simulated year 2015 allows population, industry, and pollution to rise too high. Even the effective technologies operating in this scenario cannot forestall a decline, although they do manage to reverse the decline at the end of the 21st century.

SCENARIO 12

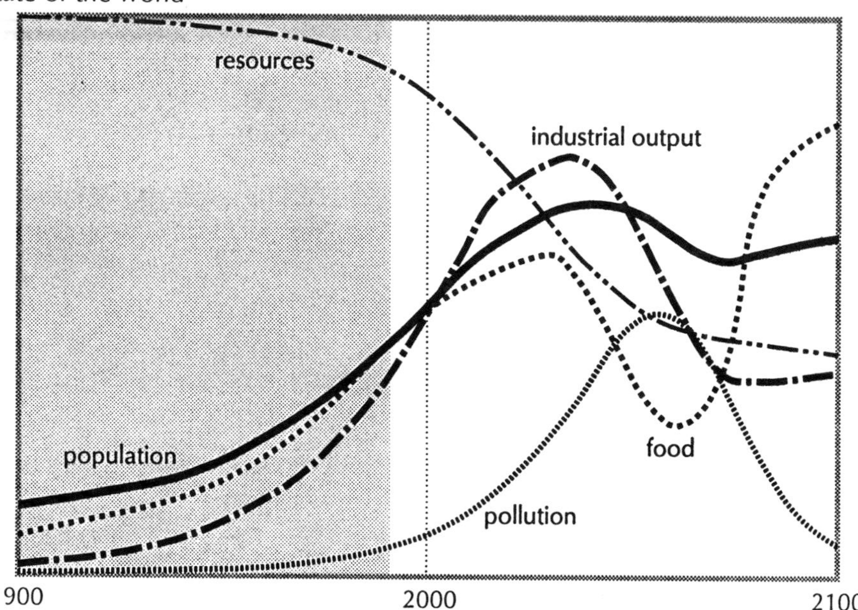

Material standard of living

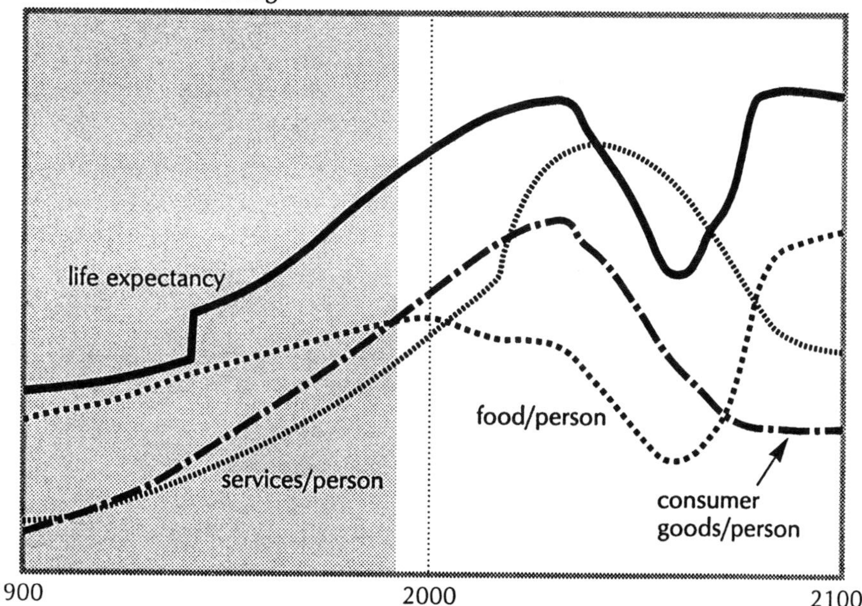

1968-$500 shortly after the year 2035 and falls slowly thereafter. Food per capita manages to hit its goal around the year 2065, but then it too begins to decline. Too much technology is necessary, too much capital is diverted to offset the damage to the environment and to attain the higher material goals. By the simulated year 2100 the per capita flows of food and industrial goods available to this more ambitious world are falling to levels lower than they were in the world of Scenario 10, which was content to set more moderate goals.

Does this run give a reliable estimate for the standard of living that would be too high for a "real world" of 8 billion people to sustain?

No. Absolutely not. The numbers are not that good. It is possible that more people actually could be supported at a higher standard of living. It could also be, given the optimistic assumptions in World3 about no war, no conflict, no corruption, and no mistakes, that Scenario 12 is much too optimistic. World3 cannot be used to fine-tune a human world seeking to find its exact upper limits. No model now available, and probably no model ever available, will permit that kind of numerical precision.

The lessons to be drawn from World3 are qualitative, not quantitative. They do not spell out an exact prediction for the future or a detailed plan for the world. But the runs shown in this chapter suggest general conclusions that are not at all recognized in the global public discourse, and that are vitally important to decisions being made (and not being made) every day. Imagine how differently the world system might behave if the following conclusions were widely known and accepted:

- A transition to a sustainable society is probably possible without reductions in either population or industrial output.

Scenario 13 EQUILIBRIUM POLICIES BUT WITH HIGHER GOALS FOR FOOD AND INDUSTRIAL OUTPUT

Using the same general policies as were implemented in Scenario 11, but with much higher demands for food and consumption places much greater stress on the global resource base. Initially the living standard is higher, but by 2100 the simulated world shows clear signs of unsustainability.

SCENARIO 13

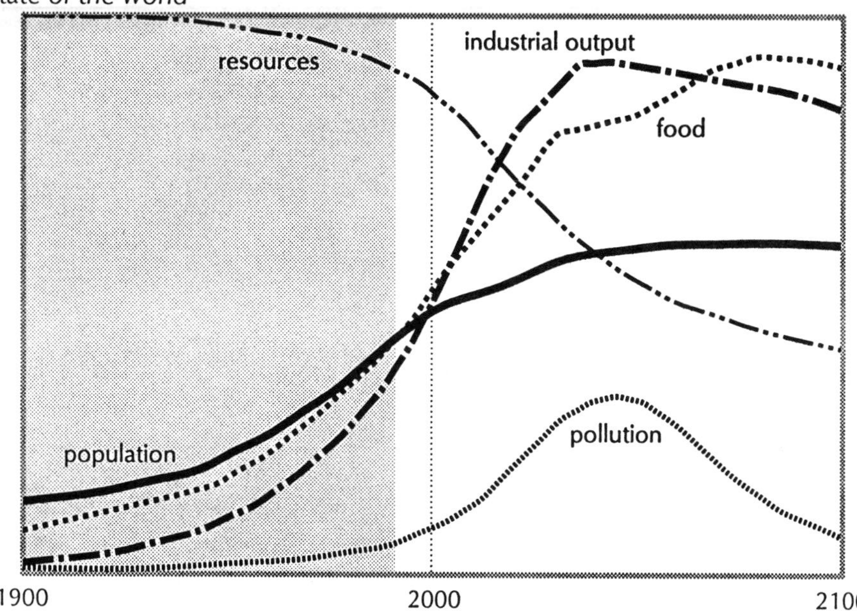

Material standard of living

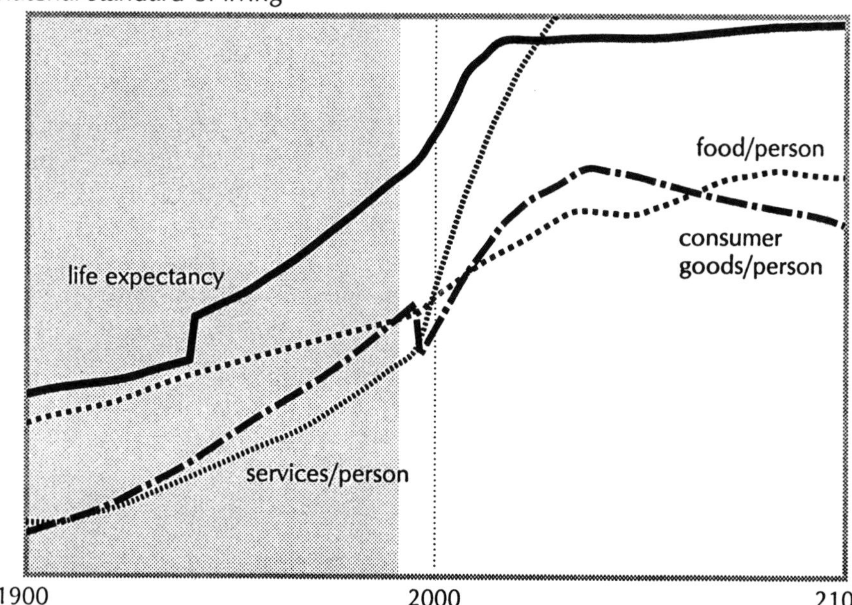

- A transition to sustainability will require, however, both deliberate social constraints on further population and industrial growth and significant improvements in the technical efficiency with which the earth's resources are used.
- There are many ways in which a sustainable society could be structured, many choices about numbers of people, living standards, technological investments, and allocations among industrial goods, services, food, and other material needs.
- As the earth's limits are approached, and especially as they are exceeded, there are unavoidable trade-offs between the number of people the earth can support and the material level at which each person can be supported. The exact numerical trade-offs are not knowable and they will change over time as technology, knowledge, human coping ability, and the earth's support systems change. The supportable population and living standard may move up or down. But the general implication of the trade-off will remain the same: More people means less material throughput for each person—or a higher risk of collapse.
- The longer the world economy takes to reduce its throughputs and move toward sustainability, the lower the population and material standard that will be ultimately supportable. At some point delay means collapse.
- The higher the society sets its targets for material standard of living, the greater its risk of exceeding and eroding its limits.

According to our computer model, our mental models, our knowledge of the data, and our experience of the "real world," there is no time to waste in easing down below the limits and structuring the information system toward sustainability. Putting off the reduction of throughputs and the transition to sustainability means diminishing the options of future generations at best, and precipitating a collapse at worst. There is no time to wait for unmistakable signals, recognizable by everyone everywhere, that force an end to growth. Given the delays in the system, by the time those signals appear, it will be too late to avoid collapse.

There is no time to waste, and there is also no reason to waste time.

Sustainability is a new idea to many people. It may be difficult to understand. But all over the world there are people who have entered into the exercise of imagining a sustainable world. They can see it as a world to move toward not reluctantly, not with a sense of sacrifice or regret, but joyfully. It could be a very much better world than the one we live in today.

The Sustainable Society

There are many ways to define sustainability. The simplest definition is: A sustainable society is one that can persist over generations, one that is far-seeing enough, flexible enough, and wise enough not to undermine either its physical or its social systems of support.

The World Commission on Environment and Development put that definition into memorable words: A sustainable society is one that "meets the needs of the present without compromising the ability of future generations to meet their own needs."[2]

From a systems point of view a sustainable society is one that has in place informational, social, and institutional mechanisms to keep in check the positive feedback loops that cause exponential population and capital growth. That means that birth rates roughly equal death rates, and investment rates roughly equal depreciation rates, unless and until technical changes and social decisions justify a considered and controlled change in the levels of population or capital. In order to be socially sustainable the combination of population, capital, and technology in the society would have to be configured so that the material living standard is adequate and secure for everyone. In order to be physically sustainable the society's material and energy throughputs would have to meet economist Herman Daly's three conditions:[3]

- Its rates of use of renewable resources do not exceed their rates of regeneration.
- Its rates of use of nonrenewable resources do not exceed the rate at which sustainable renewable substitutes are developed.
- Its rates of pollution emission do not exceed the assimilative capacity of the environment.

Whatever such a society would be like in detail, it could hardly be more different from the one in which most people now live. The collective human imagination is strongly imprinted by its recent experience either of poverty or of rapid material growth and of determined efforts to maintain that growth at all costs. Therefore many mental models are too full of growth-dominated notions to allow imagining a sustainable society. Before we can elaborate on what sustainability *could* be, we need to state what it *need not* be.

Sustainability does not mean no growth. A society fixated on perpetual growth tends to hear any criticism of growth as a total negation. But as Aurelio Peccei, founder of The Club of Rome, pointed out, that reaction just substitutes one oversimplification for another:

> All those who had helped to shatter the myth of growth . . . were ridiculed and figuratively hanged, drawn, and quartered by the loyal defenders of the sacred cow of growth. Some of those . . . accuse the [*Limits to Growth*] report . . . of advocating ZERO GROWTH. Clearly, such people have not understood anything, either about the Club of Rome, or about growth. The notion of zero growth is so primitive—as, for that matter, is that of infinite growth—and so imprecise, that it is conceptual nonsense to talk of it in a living, dynamic society.[4]

A sustainable society would be interested in qualitative development, not physical expansion. It would use material growth as a considered tool, not as a perpetual mandate. It would be neither for nor against growth, rather it would begin to discriminate kinds of growth and purposes for growth. Before this society would decide on any specific growth proposal, it would ask what the growth is for, and who would benefit, and what it would cost, and how long it would last, and whether it could be accommodated by the sources and sinks of the planet. A sustainable society would apply its values and its best knowledge of the earth's limits to choose only those kinds of growth that would actually serve social goals and enhance sustainability. And when any physical growth had accomplished its purposes, it would be brought to a stop.

A sustainable society would not freeze into permanence the current inequitable patterns of distribution. It would certainly not permit the persistence of poverty. To do so would not be sustainable for two reasons. First the poor would not and should not stand for it. Second, keeping any part of the population in poverty would not, except under dire coercive measures, allow the population to stabilize. For both moral and practical reasons any sustainable society must provide material sufficiency and security for all. To get to sustainability from here, the remaining material growth possible—whatever space there is for more resource use and pollution emissions, plus whatever space is freed up by higher efficiencies and lifestyle moderations on the part of the rich—would logically be allocated to those who need it most.

A sustainable state would not be the society of despondency and stagnancy, high unemployment and bankruptcy that current market systems experience when their growth is interrupted. The difference between a sustainable society and a present-day economic recession is like the difference between stopping an automobile purposely with the brakes and stopping it by crashing into a brick wall. When the present economy overshoots, it turns around too fast and too unexpectedly for people or enterprises to retrain, relocate, readjust. A transition to sustainability could take place slowly enough and with enough forewarning so that people and businesses could find their proper place in the new society.

There is no reason why a sustainable society need be technically or culturally primitive. Freed from both material anxiety and material greed, human society would have enormous possibilities for the expansion of human creativity in constructive directions. Without the high costs of growth for both human society and the environment, both technology and culture could bloom. John Stuart Mill, one of the first (and last) economists to take seriously the idea of an economy consistent with the limits of the earth, saw that what he called a "stationary state" could support an ever-evolving and improving society. More than a hundred years ago he wrote:

> I cannot . . . regard the stationary state of capital and wealth with the unaffected aversion so generally manifested towards it by political economists of the old school. I am inclined to believe that it would be, on the whole, a very considerable improvement on our present condition. I confess I am not charmed with the ideal of life

> held out by those who think that the normal state of human beings is that of struggling to get on; that the trampling, crushing, elbowing, and treading on each other's heels . . . are the most desirable lot of humankind. . . . It is scarcely necessary to remark that a stationary condition of capital and population implies no stationary state of human improvement. There would be as much scope as ever for all kinds of mental culture and moral and social progress; as much room for improving the Art of Living, and much more likelihood of its being improved.[5]

A sustainable world would not and could not be a rigid one, with population or production or anything else held pathologically constant. One of the strangest assumptions of present-day mental models is the widespread idea that a world of moderation must be a world of strict, centralized, government control. We don't believe that kind of control is possible, desirable, or necessary. A sustainable world would need to have rules, laws, standards, boundaries, and social agreements, of course, as does every human culture. Some of the rules for sustainability would be different from the rules people are used to now. Some of the necessary controls are already coming into being, as, for example, in the international ozone agreement.

But rules for sustainability, like every workable social rule, would not remove important freedoms; they would create them or protect them against those who would destroy them. A ban on bank-robbing inhibits the freedom of the thief in order to assure everyone's freedom to deposit and withdraw their money safely. A ban on overuse of a resource or on generation of pollution serves a similar purpose.

It doesn't take much imagining to come up with a minimum set of social structures—feedback loops that carry new information about costs, consequences, and sanctions—that would keep a society sustainable, allow evolution, fluctuation, creativity and change, and permit many more freedoms than would ever be possible in a world that continues to crowd against its limits.

Some people think that a sustainable society would have to stop using nonrenewable resources, since their use is by definition unsustainable. That idea is an overly rigid interpretation of what it means to be sustainable. Certainly a sustainable society would use gifts from the earth's crust more thoughtfully and efficiently than the present world does. It would price them properly and keep more of them available for future generations. But there is no reason not to use them, as long as their use meets the criterion of sustainability already defined, namely that renewable substitutes should be developed, so that no future society finds itself built around the use of a resource that is suddenly no longer available or affordable.

There is also no reason for a sustainable society to be uniform. Diversity is both a cause of and a result of sustainability in nature, and it would be in human society as well. Most people envision a sustainable world as decentralized, with boundary conditions keeping each locality from threatening the viability of another or of the earth as a whole. Cultural variety and local autonomy could be greater, not less, in such a world.

There is no reason for a sustainable society to be undemocratic, or boring, or unchallenging. Some games that amuse and consume people today, such as arms races and the accumulation of unlimited amounts of wealth, would no longer be played. But there still would be games, challenges, problems to solve, ways for people to prove themselves, to serve each other, to realize their abilities, and to live good lives, perhaps more satisfying lives than any that are possible today.

That was a long list of what a sustainable society is not. In the process of spelling it out, we have also, by contrast, indicated what we think a sustainable society could be. But the details of that society will not be worked out by one bunch of computer modelers; it will require the ideas, visions, and talents of billions of people.

From the structural analysis of the world system we have described in this book, we can contribute only a simple set of general guidelines for restructuring the world system toward sustainability. We list the guidelines below. Each one can be worked out in hundreds of specific ways at all levels from households to communities to nations to the world as a whole. Other people will see better than we can how to implement these changes in their own lives and cultures and political systems. Any step in any of these directions is a step toward sustainability.

- *Improve the signals.* Learn more about and monitor both the welfare of the human population and the condition of local and planetary sources and sinks. Inform governments and the public as continuously and promptly about environmental conditions as about economic conditions. Include real environmental costs in economic prices; recast economic indicators like the GNP so that they do not confuse costs with benefits, or throughput with welfare, or the depreciation of natural capital with income.[6]

- *Speed up response times.* Look actively for signals that indicate when the environment is stressed. Decide in advance what to do if problems appear (if possible, forecast them before they appear) and have in place the institutional and technical arrangements necessary to act effectively. Educate for flexibility and creativity, for critical thinking and the ability to redesign both physical and social systems. Computer modeling can help with this step, but more important would be general education in systems thinking.

- *Minimize the use of nonrenewable resources.* Fossil fuels, fossil groundwaters, and minerals should be used only with the greatest possible efficiency, recycled when possible (fuels can't be recycled, but minerals and water can), and consumed only as part of a deliberate transition to renewable resources.

- *Prevent the erosion of renewable resources.* The productivity of soils, surface waters, rechargeable groundwaters, and all living things, including forests, fish, game should be protected and, as far as possible, restored and enhanced. These resources should only be harvested at the rate they can regenerate themselves. That requires information about their regeneration rates, and strong social sanctions or economic inducements against their overuse

- *Use all resources with maximum efficiency.* The more human welfare can be obtained with the less throughput, the better the quality of life can be while remaining below the limits. Great efficiency gains are both technically possible and economically favorable. Higher efficiency will be essential, if current and future world populations are to be supported without inducing a collapse.

- *Slow and eventually stop exponential growth of population and physical capital.* There are real limits to the extent that the first five items on this list can be pursued. Therefore this last item is essential. It involves institutional and philosophical change and social innovation. It requires defining levels of population and industrial output that are desirable and sustainable. It calls for goals defined around the idea of development rather than growth. It asks, simply but profoundly, for a vision of the purpose of human existence that does not require constant physical expansion.

We can expand on this last, most daunting, but most important step toward sustainability by pointing to the pressing problems that underlie much of the psychological and cultural commitment to growth: poverty, unemployment, and unmet nonmaterial needs. Growth as presently structured is in fact not solving these problems, or is solving them far too slowly and inefficiently. Until better solutions are in sight, however, society will never let go of its addiction to growth. These are the three areas where completely new thinking is most urgently needed.

- *Poverty.* "Sharing" is a forbidden word in political discourse, probably because of the deep fear that real equity would mean not enough for anyone. "Sufficiency" and "solidarity" are concepts that can help structure new approaches to ending poverty. Everyone needs assurance that sufficiency is possible and that there is a high social commitment to ensure it. And everyone needs to understand that the world is tied together both ecologically and economically. We are all in this overshoot together. There is enough to go around, if we manage well. If we don't manage well, no one will escape the consequences.

- *Unemployment.* Human beings need to work, to have the satisfaction of personal productivity, and to be accepted as responsible members of their society. That need should be not be left unfulfilled, and it should not be filled by degrading or harmful work. At the same time, employment should not be a requirement for the ability to subsist. Considerable creativity is necessary here to create an economic system that uses and supports the contributions that all people are able and willing to make, that shares work and leisure equitably, and that does not abandon people who for reasons temporary or permanent cannot work.

- *Unmet nonmaterial needs.* People don't need enormous cars; they need respect. They don't need closetsful of clothes; they need to feel attractive and they need excitement and variety and beauty. People don't need electronic entertainment; they need something worthwhile to do with their lives. And so forth. People need identity, community, challenge, acknowledgment, love, joy. To try to fill these needs with material things is to set up an unquenchable appetite for false solutions to real and never-satisfied problems. The resulting psychological emptiness is one of the major forces behind the desire for material growth. A society that can admit and articulate its nonmaterial needs and find nonmaterial ways to satisfy them would require much lower material and energy throughputs and would provide much higher levels of human fulfillment.

How, in practice, can anyone attack these problems? How can the world evolve a social *system* that solves them? That is the real arena for creativity and choice. It is necessary for the present generation not only to bring itself below the earth's limits but to restructure its inner and outer worlds. That process will touch every arena of life. It will require every kind of human talent. It will need not only technical and entrepreneurial innovation, but also communal, social, political, artistic, and spiritual innovation. Lewis Mumford recognized fifty years ago not only the magnitude of the task, but also the fact that it is a particularly *human* task, one that will challenge and develop the *humanity*—in the most noble sense of that word—of everyone.

> An age of expansion is giving place to an age of equilibrium. The achievement of this equilibrium is the task of the next few centuries. . . . The theme for the new period will be neither arms and the man nor machines and the man: its theme will be the resurgence of life, the displacement of the mechanical by the organic, and the re-establishment of the person as the ultimate term of all human effort. Cultivation, humanization, co-operation, symbiosis: these are the watchwords of the new world-enveloping culture. Every department of life will record this change: it will affect the task of education and the procedures of science no less than the organization of industrial enterprises, the planning of cities, the development of regions, the interchange of world resources.[7]

The necessity to take the industrial world of growth to its next stage of evolution is not a disaster, it is an opportunity. How to seize the opportunity, how to bring into being a sustainable world that is not only functional but desirable is a question about leadership and ethics and vision and courage. Those are properties not of technologies, markets, government, corporations, or computer models but of the human heart and soul. To speak of them the authors need a chapter break here, to take off their computer modeling hats and put away their scientists' white coats and to reappear as plain human beings.

chapter 8

Overshoot but Not Collapse

Can we move nations and people in the direction of sustainability? Such a move would be a modification of society comparable in scale to only two other changes: the Agricultural Revolution of the late Neolithic and the Industrial Revolution of the past two centuries. Those revolutions were gradual, spontaneous, and largely unconscious. This one will have to be a fully conscious operation, guided by the best foresight that science can provide. . . . If we actually do it, the undertaking will be absolutely unique in humanity's stay on the Earth.

William D. Ruckelshaus[1]

We have personally been writing about, talking about, and working toward sustainability for more than twenty years. We have had the privilege of knowing thousands of people in every part of the world who are working in the same direction, in their own ways, with their own talents, in their own countries. When we act at the official, institutional level and when we listen to political leaders, we often feel frustrated. When we work with individuals outside the boundaries of institutions, we usually feel encouraged.

Everywhere we find people who care about the Earth, about other people, and about the welfare of their children and grandchildren. They recognize the human misery and the environmental degradation already apparent in the world, and they question whether current policies that promote growth can make things better. They are willing to work for a sustainable society, if they could only believe that their efforts would make a useful difference. They ask: What can I do? What can governments do? What can corporations do? What can schools, religions, media do? What can citizens, producers, consumers, parents do?

We think that committed experimentation guided by those questions is more important than any particular answer, though answers abound. There are "fifty simple things you can do to save the planet." Buy an energy-efficient car, recycle your bottles and cans, vote knowledgeably in political elections–if you are one of the people in the world who have cars, bottles, cans, or elections. There are also not-so-simple things to do: work out your own conserving lifestyle; have at most two children; work with love and partnership to help one family lift itself out of poverty; earn your living in "right livelihood"; care responsibly for one piece of land; do whatever you can to avoid supporting systems that oppress people or abuse the earth.

All these actions will help. They are all necessary. And, of course, they are not enough. We are talking about a revolution here, not in the political sense, like the French Revolution, but in the much more profound sense of the Agricultural or Industrial Revolution.[2] Recycling bottles and cans is a good idea, but by itself it will not bring the world to that kind of revolution.

What will? In search of an answer to that question, we have found it helpful to try to understand the first two great revolutions, insofar as historians can reconstruct them.

The First Two Revolutions: Agriculture and Industry

About 8000 years ago, the human population, after eons of slow accumulation, had reached the enormous (for the time) number of about 10 million. These people lived as nomadic hunter-gatherers, but their numbers had begun to overwhelm the useful plants and game that had until then been abundant all around them. To adapt to the problem of disappearing wild resources they did two things. Some of them intensified their migratory lifestyle. They starting moving out of their ancestral homes of Africa and the Middle East and populating the rest of the game-rich world.

Other people started domesticating animals, cultivating plants, and as a consequence *staying in one place.* That was a totally new idea.

Simply by staying put, the proto-farmers altered the face of the planet and the thoughts of humankind in ways they could never have foreseen.

For example, for the first time it made sense to own land. Furthermore, people who didn't have to carry all their possessions on their backs could accumulate things. Some people could accumulate more than others. The ideas of wealth, inheritance, trade, money, and power were born. Some people could live on excess food produced by others and become full-time potters, toolmakers, musicians, scribes, priests, soldiers, or kings. Thus arose, for better or worse, cities, experts, entertainers, armies, and bureaucrats.

As its inheritors, we think of the Agricultural Revolution as a great step forward. At the time, however, it was apparently a mixed blessing. Many anthropologists think that agriculture was not a better way of life, but a necessary one to accommodate increasing human populations. Settled farmers got more food from an acre of land than hunter-gatherers did, but the food was of lower nutritional quality and less variety, and it required much more work. Farmers became vulnerable in ways nomads never had been to weather, disease, pests, invasion by outsiders, and oppression from their own emerging ruling class. Since settled people did not move away from their own wastes, they experienced humankind's first chronic pollution.

Agriculture was a successful response to wildlife scarcity. It permitted continued slow population growth, which added up over centuries to an enormous increase, from about 10 million people to about 800 million by 1750. By that time the larger population had created new scarcities, especially in land and energy. Another revolution was necessary.

The Industrial Revolution began in England with the substitution of abundant coal for vanishing trees. The use of coal raised immediate practical problems of earth moving, mine construction, water pumping, transport, and controlled combustion. It required greater concentrations of labor around the mines and mills, and it required the elevation of science and technology to prominent positions in human society.

Again everything changed in ways that no one could have imagined. Coal led to steam engines. Machines, not land, became the central means of production. Feudalism therefore gave way to capitalism and to capitalism's dissenting offshoot, communism. Roads, railroads, factories, smokestacks appeared everywhere. Cities swelled. Again the change was a mixed blessing. Factory labor was even harder and more demeaning than farm labor. The environment near the new factories turned unspeakably filthy. The standard of living for most people in the industrial work force was far below that of a yeoman farmer. But work in a factory was better than starving on the crowded land.

It is hard for people alive today to appreciate how profoundly the Industrial Revolution changed human thought, because we still think its thought. Historian Donald Worster has described the philosophical impact of industrialism perhaps as well as any of its inheritors and practitioners can:

> The capitalists . . . promised that, through the technological domination of the earth, they could deliver a more fair, rational, efficient and productive life for everyone. . . . Their method was simply to free individual enterprise from the bonds of traditional hierarchy and community, whether the bondage derived from other humans or the earth. That meant teaching everyone to treat the earth, as well as each other, with a frank, energetic, self-assertiveness. . . . People must . . . think constantly in terms of making money. They must regard everything around them—the land, its natural resources, their own labor—as potential commodities that might fetch a profit in the market. They must demand the right to produce, buy, and sell those commodities without outside regulation or interference. . . . As wants multiplied, as markets grew more and more far-flung, the bond between humans and the rest of nature was reduced to the barest instrumentalism.[3]

That bare instrumentalism led to great material productivity and a world that now supports, partially anyway, more than 5 billion people. The far-flung markets led to environmental exploitation from the poles to the tropics, from the mountaintops to the ocean depths. The success of the Industrial Revolution, like the more limited successes of hunting-gathering and of agriculture, eventually led to further scarcities, not only of game, not only of land, not only of fuels and metals, but of the absorptive capacity of the environment.

Therefore it has created the necessity for another revolution.

The Next Revolution: Sustainability

It is as impossible for anyone now to describe the world that could evolve from a sustainability revolution as it would have been for the farmers of 6000 BC to foresee present-day Iowa, or for an English coal miner of 1750 to imagine a Toyota assembly line. The most anyone can say is that, like the other great revolutions, a sustainability revolution could lead to enormous gains and losses. It too could change the face of the land and the foundations of human self-definitions, institutions, and cultures. Like the other revolutions, it will take centuries to develop fully—though we believe it is already underway and that its next steps need to be taken with urgency, so that a revolution is possible instead of a collapse.

Of course no one knows how to bring about a sustainability revolution. There is not and will never be a checkoff list: "To accomplish a global revolution, follow the twenty steps below." Like the revolutions that came before, this one can't be planned or dictated. It won't follow a list of fiats from a government or from computer modelers. The Sustainability Revolution, if it happens, will be organic and evolutionary. It will arise from the visions, insights, experiments, and actions of billions of people. The burden of making it happen is not on the shoulders of any one person or group. No identifiable person or group will get the credit, though some may get some blame. And everyone can contribute.

Our systems training and our own work in the world have brought home to us two properties of complex systems that are important to the sort of thoroughgoing revolution we are discussing here.

First, information is the key to transformation. That does not necessarily mean more information, better statistics, bigger databases. It means information flowing in new ways, to new recipients, carrying new content, and suggesting new rules and goals (rules and goals are themselves information). With different information structures, the system will inevitably behave differently. The policy of *glasnost,* for example, the simple opening of information channels that had long been closed, guaranteed the rapid transformation of Eastern Europe. The old system had been held in place by tight control of information.

Letting go of that control necessitated some sort of restructuring (turbulent and unpredictable, but inevitable) toward a new system consistent with the new information.

Second, systems strongly resist changes in their information flows, most especially in their rules and goals. An existing system can constrain almost entirely the attempts of an individual to operate by different rules or to attain different goals than those sanctioned by the system. However, only individuals, by perceiving the need for new information, rules, and goals, communicating about them, and trying them out, can make the changes that transform systems.

For example, we have learned the hard way that it is difficult to live a life of material moderation within a social system that expects, exhorts, values, and rewards consumption. But an individual can move a long way in the direction of moderation. It is not easy to use energy efficiently in an economy that produces energy-inefficient products. But one can search out or, if necessary, invent more efficient ways of doing things. Above all, it is difficult to put forth new information in a system that is structured to hear and process only old information. Just try, sometime, to question in public the value of more physical growth, or even to make a distinction between growth and development, and you will see what we mean. It takes courage and clarity to communicate information that challenges the structure of an established system. But it can be done.

In our own search for ways to encourage the peaceful restructuring of a system that naturally resists its own transformation, we have tried many tools. The most obvious ones are displayed throughout this book—rational analysis, data, systems thinking, computer modeling, and the clearest words we are capable of finding to express new information and new models. Those are tools that anyone trained like us in science and economics would automatically grasp. Like recycling, they are useful, necessary, and not enough.

We don't know what will be enough. But we would like to conclude this book by mentioning five other tools we have found helpful, not as *the* ways to work toward sustainability, but as *some* ways that have been useful to us. We are a bit hesitant to discuss them because we are not experts in their use and because they require the use of words that do not come easily from the mouths or word processors of scientists. They are considered too "soft" to be taken seriously in the cynical public arena. They are: visioning, networking, truth-telling, learning, and loving.

The transition to a sustainable society might be helped by the simple use of words like these more often, with sincerity and without apology, in the information streams of the world.

Visioning

Visioning means imagining, at first generally and then with increasing specificity, what you really want. That is, *what you really want,* not what someone has taught you to want, and not what you have learned to be willing to settle for. Visioning means taking off all the constraints of assumed "feasibility," of disbelief and past disappointments, and letting your mind dwell upon its most noble, uplifting, treasured dreams.

Some people, especially young people, engage in visioning with enthusiasm and ease. Some people find the exercise of visioning painful, because a glowing picture of what *could be* makes what *is* all the more intolerable. Some people would never admit their visions, for fear of being thought impractical or "unrealistic." They would find this paragraph uncomfortable to read, if they were willing to read it at all. And some people have been so crushed by their experience of the world that they can only stand ready to explain why any vision is impossible. That's fine; they are needed too. Vision needs to be balanced with skepticism.

We should say immediately for the sake of the skeptics that we do not believe it is possible for the world to envision its way to a sustainable future. Vision without action is useless. But action without vision does not know where to go or why to go there. Vision is absolutely necessary to guide and motivate action. More than that, vision, when widely shared and firmly kept in sight, *brings into being new systems.*

We mean that literally. Within the physical limits of space, time, materials, and energy, visionary human intentions can bring forth not only new information, new behavior, new knowledge, and new technology, but eventually new social institutions, new physical structures, and new powers within human beings. Ralph Waldo Emerson recognized this strange truth 150 years ago:

> Every nation and every man instantly surround themselves with a material apparatus which exactly corresponds to their moral state, or their state of thought. Observe how every truth and every error, each a thought of some man's mind, clothes itself with societies, houses, cities, language, ceremonies, newspapers. Observe the ideas of the present day . . . see how each of these abstractions has embodied itself in an imposing apparatus in the community, and how timber, brick, lime, and stone have flown into convenient shape, obedient to the master idea reigning in the minds of many persons. . . . It follows, of course, that the least change in the man will change his circumstances; the least enlargement of ideas, the least mitigation of his feelings in respect to other men . . . would cause the most striking changes of external things.[4]

A sustainable world can never come into being if it cannot be envisioned. The vision must be built up from the contributions of many people before it is complete and compelling. As a way of encouraging others to join in the process of visioning, we'll list here some of what we see, when we let ourselves imagine a sustainable society that we would like to live in.[5] This is by no means a definitive list or a complete vision. We include it only to invite you to develop and enlarge it.

- Sustainability, efficiency, sufficiency, justice, equity, and community as high social values.
- Leaders who are honest, respectful, and more interested in doing their jobs than in keeping their jobs. (Remember, this is a vision, not what we have come to expect.)
- Material sufficiency and security for all. Therefore, by spontaneous choice as well as by communal norms, low death rates, low birth rates, and stable populations.
- Work that dignifies people instead of demeaning them. Some way of providing incentives for people to give of their best to society and to be rewarded for doing so, while still ensuring that people will be provided for sufficiently under any circumstances.
- An economy that is a means, not an end, one that serves the wel-

fare of the human community and the environment, rather than demanding that the community and environment serve it.[6]

- Efficient, renewable energy systems; efficient, cyclic materials systems.

- Technical design that reduces pollution and waste to a minimum, and social agreement not to produce pollution or waste that nature can't handle.

- Regenerative agriculture that builds soils, uses natural mechanisms to restore nutrients and control pests, and produces abundant, uncontaminated food.

- Preservation of ecosystems in their variety, with human cultures living in harmony with those ecosystems—therefore high diversity of both nature and culture, and human tolerance and appreciation for that diversity.

- Flexibility, innovation (social as well as technical), and intellectual challenge. A flourishing of science, a continuous enlargement of human knowledge.

- Greater understanding of whole systems as an essential part of each person's education.

- Decentralization of economic power, political influence, and scientific expertise.

- Political structures that permit a balance between short-term and long-term considerations. Some way of exerting political pressure on behalf of the grandchildren.

- High skills on the part of citizens and governments in the arts of nonviolent conflict resolution.

- Print and broadcast media that reflect the world's diversity and at the same time bind together the cultures of the world with relevant, accurate, timely, unbiased, and intelligent information, set into its historic and whole-system context.

- Reasons for living and for thinking well of oneself that do not require the accumulation of material things.

Networking

We could not do our work without networks. Most of the networks we belong to are informal. They have small budgets, if any, and few of them appear on rosters of world organizations. They are almost invisible, but their effects are not negligible. Informal networks carry information in the same way as formal institutions do, and often more effectively. They are the natural home of new information, and out of them can evolve new system structures.

Some of the networks important to us are very local, some are international. They are simply collections of people who stay in touch, who pass around data and tools and ideas and, most important of all, encouragement. One of the important purposes of a network is simply to remind its members that they are not alone.

A network is by definition nonhierarchical. It is a web of connections among equals. What holds it together is not force, obligation, material incentive, or social contract, but rather shared values and the understanding that some tasks can be accomplished together that could never be accomplished separately.

We know of networks of farmers who are exploring organic methods and sharing their experience. There are networks of environmental journalists, of "green" entrepreneurs, of computer modelers, of game designers, land trusts, consumer cooperatives. There are thousands and thousands of networks. They spring up naturally as human beings with common purposes find each other. Some networks become so big and busy and essential that they evolve into formal organizations with offices and budgets, but most just come and go as needed.

Networks dedicated to sustainability seem to be forming most actively at the local and the global levels. New organizations at these levels may be especially needed to create a sustainable society that harmonizes with local ecosystems while keeping itself below global limits. About local networks we can say little here; our localities are different from yours. One role of local networks is to help reestablish the sense of community and of relationship to place that has been largely lost since the Industrial Revolution.

When it comes to global networks, we would like to make a plea that they be truly global. The means of participation in international information streams are as badly distributed among the world's people as are the means of production. There are more telephones in Tokyo, it is said, than in all of Africa. That must be even more true of computers, fax machines, airline connections, and invitations to international meetings.

It could be argued that Africa and other underrepresented parts of the world have urgent needs for many things other than telephones and fax machines. We suggest that those needs cannot be effectively expressed in the world, nor can the world benefit from the contributions of underrepresented peoples unless their voices can become part of the global conversation. Some of the greatest gains in material and energy efficiency have come in the design of communications equipment. It should be possible within the throughput limits of the earth for everyone to have the opportunity for global as well as local networking.

If you see a part of the sustainability revolution that interests you, you can find or form a network of others who share that interest. The network will help you discover where to go for information, what publications and tools are available, where to find administrative and financial support, and who can join you for specific tasks. The right network will not only help you learn but will allow you to pass your learning on to others.

Truth-telling

No, we are no more certain of the truth than anyone is. But we often recognize an untruth when we hear one, coming from our own mouths or those of others, and most particularly coming from advertisers and from political leaders. Many of those untruths are deliberate, understood as such by both speakers and listeners. They are put forth to manipulate, lull, or entice, to postpone action, to justify self-serving action, to gain or preserve power, or to deny an uncomfortable reality.

Lies distort the information stream. A system cannot function, especially in time of peril, if its information stream is confused and distorted. One of the most important tenets of systems theory, for reasons we hope we have made clear in this book, is that information should not be deliberately distorted, delayed, or sequestered.

"All of humanity is in peril," said Buckminster Fuller, "if each one of us does not dare, now and henceforth, always to tell only the truth and all the truth, and to do so promptly–right now."[7] Whenever you speak in public, even to a public of one other person, you can counter a lie or affirm a truth, as best you can see truth. You can deny the idea that having more things makes one a better person, or you can endorse it. You can question the idea that growth for the rich will help the poor, or you can accept it. The more you can use your voice to counter misinformation, the more manageable your society will ultimately be.

Here are some common biases and simplifications, verbal traps, untruths that we have run into frequently in discussing limits to growth. We think they need to be pointed out and avoided, if there is ever to be clear thinking about the human economy and its relationship to the earth.

- *Not*: A warning about the future is a prediction of doom.
  But: A warning about the future is a recommendation to follow a different path.

- *Not*: The environment is a luxury or a competing demand or a commodity that people will buy when they can afford it.
  But: The environment is the source of all life and every economy.

- *Not*: Change is sacrifice.
  But: Change is challenge and it is necessary.

- *Not*: Stopping growth will lock the poor in their poverty.
  But: Present patterns of growth are locking the poor into poverty; they need growth that is specifically geared to serve their needs.

- *Not*: Everyone should be brought up to the material level of the richest countries.
  But: All material human needs should be met materially and all nonmaterial human needs met nonmaterially.

- *Not*: All growth is good, without question, discrimination, or investigation.
  And not: All growth is bad.

But: What is needed is not growth, but development. Insofar as development requires physical expansion, it should be equitable, affordable, and sustainable.

- *Not*: Technology will solve all problems, or technology does nothing but cause problems.
  But: What technologies will reduce throughput, increase efficiency, enhance resources, improve signals, end poverty, and how can society encourage them?
  And: What can we bring to our problems as human beings, beyond our ability to produce technology?

- *Not*: The market system will automatically bring us the future we want.
  But: How do we use the market system, along with many other organizational devices, to bring us the future we want?

- *Not*: Industry is the cause of all problems, or the cure.
  Nor: Government is the cause or the cure.
  Nor: Environmentalists are the cause or the cure.
  Nor: Any other group (economists come to mind) is the cause or cure.
  But: All people and institutions play their role within the large system structure. In a system that is structured for overshoot, all players will deliberately or inadvertently contribute to that overshoot. In a system that is structured for sustainability, industries, governments, environmentalists, and most especially economists will play essential roles in contributing to sustainability.

- *Not*: Unrelieved pessimism.
  And not: Sappy optimism.
  But: The resolve to discover and tell the truth about the successes and the failures of the present and the potentials and the obstacles in the future.
  And above all: The courage to admit and bear the pain of the present world, while keeping a steady eye on a vision of a better future.

- *Not*: The World3 model, or any other model, is right or wrong.

> *But*: All models, including the ones in our heads, are a little right, much too simple, and mostly wrong. How do we proceed in such a way as to test our models and learn where they are right and wrong? How do we speak to each other as fellow modelers, with the appropriate mixture of skepticism and respect? How do we stop playing right/wrong games with each other, and start designing right/wrong tests for our models against the real world?

That last challenge, the sorting out and testing of models, brings us to the topic of learning.

Learning

Visioning, networking, and truth-telling are useless if they do not lead to action. There are many things to *do* to bring about a sustainable world. New farming methods have to be worked out. New kinds of businesses have to be started and old ones have to be redesigned to reduce throughputs. Land has to be restored, parks protected, energy systems transformed, international agreements reached. Laws have to be passed, and others repealed. Children have to be taught and so do adults. Films have to be made, music played, books published, people counseled, groups led.

Each person will find his or her own best role in all this doing. We wouldn't presume to prescribe that role for anyone but ourselves. But we would make one suggestion about *how* to do whatever you do. Do it humbly. Do it not as a declaration of policy but as an experiment. Use your action, whatever it is, to learn.

The depths of human ignorance are much more profound than most humans are willing to admit. Especially at a time when the global society is coming together as a more integrated whole than it has ever been before, when that society is pressing against the dynamic limits of a wondrously complex planet, and when wholly new ways of thinking are called for, no one really knows enough. No leader, no matter how authoritative he or she pretends to be, understands the situation. No policy can be declared as The Policy to be imposed on the world.

Learning means the willingness to go slow, to try things out, and to collect information about the effects of actions, including the crucial but not always welcome information that an action or policy is not working. One can't learn without making mistakes, telling the truth about them, and moving on. Learning means exploring a new path with vigor and courage, being open to other peoples' explorations of other paths, and being willing to switch paths if new evidence suggests that another one leads more efficiently or directly to the goal.

The world's leaders have lost both the habit of learning and the freedom to learn. Somehow a cultural system has evolved that assigns most people to the role of followers, who expect leaders to have all the answers, and assigns a few people to the role of leaders, who pretend they have all the answers. This perverse system does not allow the development of either leadership capability in the people or learning capability in the leaders.

It's time to do some truth-telling. The world's leaders do not know any better than anyone else how to bring about a sustainable society; most of them don't even know it's necessary to do so. A sustainability revolution requires each person to act as a learning leader at some level, from family to community to nation to the world. And it requires each of us to support leaders at all levels in their learning by creating an environment that permits them to admit uncertainty, conduct experiments, and acknowledge mistakes.

No one can be free to learn without patience and forgiveness. But in a condition of overshoot, with possible collapse on the horizon, there is not much time for patience and forgiveness; there is a need for action, determination, courage, and accountability. Finding the right balance between the apparent opposites of urgency and patience, accountability and forgiveness is a task that requires compassion, humility, clear-headedness, and honesty.

In the quest for a sustainable world, it doesn't take long before even the most hard-boiled, rational, and practical persons, even those who have not been trained in the language of humanism, begin to speak, with whatever words they can muster, of virtue, morality, wisdom, and love.

Loving

One is not allowed in the modern culture to speak about love, except in the most romantic and trivial sense of the word. Anyone who calls upon the capacity of people to practice brotherly and sisterly love is more likely to be ridiculed than to be taken seriously. The deepest difference between optimists and pessimists is their position in the debate about whether human beings are able to operate collectively from a basis of love. In a society that systematically develops in people their individualism, their competitiveness, and their cynicism, the pessimists are in the vast majority.

That pessimism is the single greatest problem of the current social system, we think, and the deepest cause of unsustainability. A culture that cannot believe in, discuss, and develop the best human qualities is one that suffers from a tragic distortion of information. "How good a society does human nature permit?" asked psychologist Abraham Maslow. "How good a human nature does society permit?"[8]

The sustainability revolution will have to be, above all, a societal transformation that permits the best of human nature rather than the worst to be expressed and nurtured. Many people have recognized that necessity and that opportunity. For example, John Maynard Keynes wrote in 1932:

> The problem of want and poverty and the economic struggle between classes and nations is nothing but a frightful muddle, a transitory and unnecessary muddle. For the Western World already has the resource and the technique, if we could create the organization to use them, capable of reducing the Economic Problem, which now absorbs our moral and material energy, to a position of secondary importance . . . . Thus the . . . day is not far off when the Economic Problem will take the back seat where it belongs, and . . . the arena of the heart and head will be occupied . . . by our real problems—the problems of life and of human relations, of creation and behaviour and religion.[9]

Aurelio Peccei, the great industrial leader who wrote constantly about problems of growth and limits, economics and environment, resources and governance, never failed to conclude that the answers to the world's problems begin with a "new humanism":

> The humanism consonant with our epoch must replace and reverse principles and norms that we have heretofore regarded as untouchable, but that have become inapplicable, or discordant with our purpose; it must encourage the rise of new value systems to redress our inner balance, and of new spiritual, ethical, philosophical, social, political, esthetic, and artistic motivations to fill the emptiness of our life; it must be capable of restoring within us . . . love, friendship, understanding, solidarity, a spirit of sacrifice, conviviality; and it must make us understand that the more closely these qualities link us to other forms of life and to our brothers and sisters everywhere in the world, the more we shall gain.[10]

It is difficult to speak of or to practice love, friendship, generosity, understanding, or solidarity within a system whose rules, goals, and information streams are geared for lesser human qualities. But we try, and we urge you to try. Be patient with yourself and others as you and they confront the difficulty of a changing world. Understand and empathize with inevitable resistance; there is some resistance, some clinging to the ways of unsustainability, within each of us. Include everyone in the new world. Everyone will be needed. Seek out and trust in the best human instincts in yourself and in everyone. Listen to the cynicism around you and pity those who believe it, but don't believe it yourself.

The world can never pass safely through the adventure of bringing itself below the limits if that adventure is not undertaken in a spirit of global partnership. Collapse cannot be avoided if people do not learn to view themselves and others with compassion. We take our stand as optimists. We think people can find that compassion within themselves if they are given the opportunity, without ridicule, to do so.

Is any change we have advocated in this book, from more resource efficiency to more human compassion, really possible? Can the world actually ease down below the limits and avoid collapse? Is there enough time? Is there enough money, technology, freedom, vision, community, responsibility, foresight, discipline, and love, on a global scale?

Of all the hypothetical questions we have posed in this book, those

Figure 8-1 TIME HORIZON OF THE WORLD3 MODEL

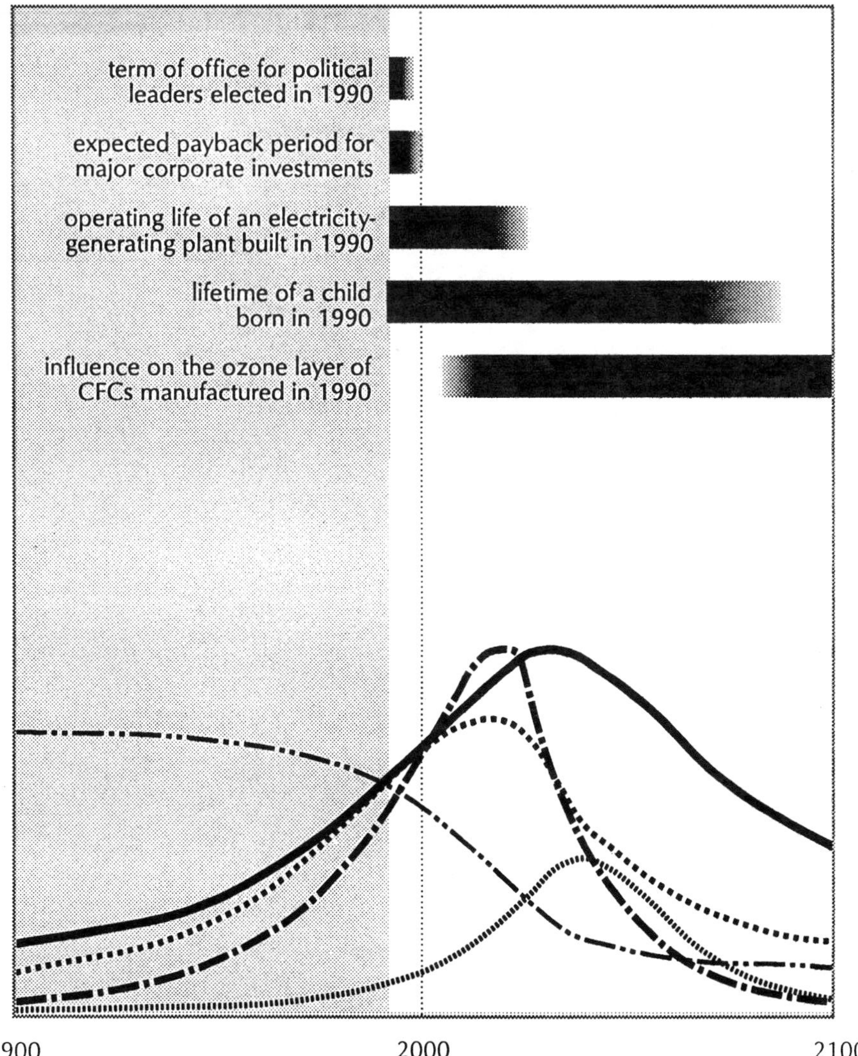

are the ones that are most unanswerable, though many people will pretend to know the answers. The ritual cheerfulness of many uninformed people, especially many world leaders, would say the questions are not even relevant; there are no meaningful limits. Many of those who are informed and who worry about the problem of overshoot are infected with the deep public cynicism that lies just under the ritual cheerfulness. They would would say that there are severe problems already, with worse ones ahead, and that there's not a chance of solving them.

Both those answers are based, of course, on mental models. The truth is that no one knows.

We have said many times in this book that the world faces not a preordained future, but a choice. The choice is between models. One model says that this finite world for all practical purposes has no limits. Choosing that model will take us even further beyond the limits and, we believe, to collapse.

Another model says that the limits are real and close, and that there is not enough time, and that people cannot be moderate or responsible or compassionate. That model is self-fulfilling. If the world chooses to believe it, the world will get to be right, and the result will also be collapse.

A third model says that the limits are real and close, and that there is just exactly enough time, with no time to waste. There is just exactly enough energy, enough material, enough money, enough environmental resilience, and enough human virtue to bring about a revolution to a better world.

That model might be wrong. All the evidence we have seen, from the world data to the global computer models, suggests that it might be right. There is no way of knowing for sure, other than to try it.

appendix

Research and Teaching with World3

When a model has reached the formal perfection of World3, and when so much effort and talent have gone into presenting its methodology in intelligible detail, its conclusions cannot be dismissed without resorting to similar methods and raising new questions to be answered by new models.

Etienne van de Walle[1]

In this book we have described basic insights and conclusions about the long-term causes and consequences of physical growth in the global system. Our team originally developed these insights through two years of research that involved designing, building, and analyzing the formal mathematical model, World3. However, judging the plausibility of our results does not require that you run the model yourself; to evaluate our conclusions the great majority of readers will only need an intuitive understanding of dynamics based on "real-world" experience with exponential change, limits, delays, and errors in perception and response.

Consequently, most people who read our report will be able to judge how much credence they place in our results without resorting to the computer. But if you wish to use World3 in your research or your teaching, you must, of course, study the details of our model. You may wish to replicate our projections, develop new scenarios that show the

Figure A-1 PERSISTENT POLLUTION

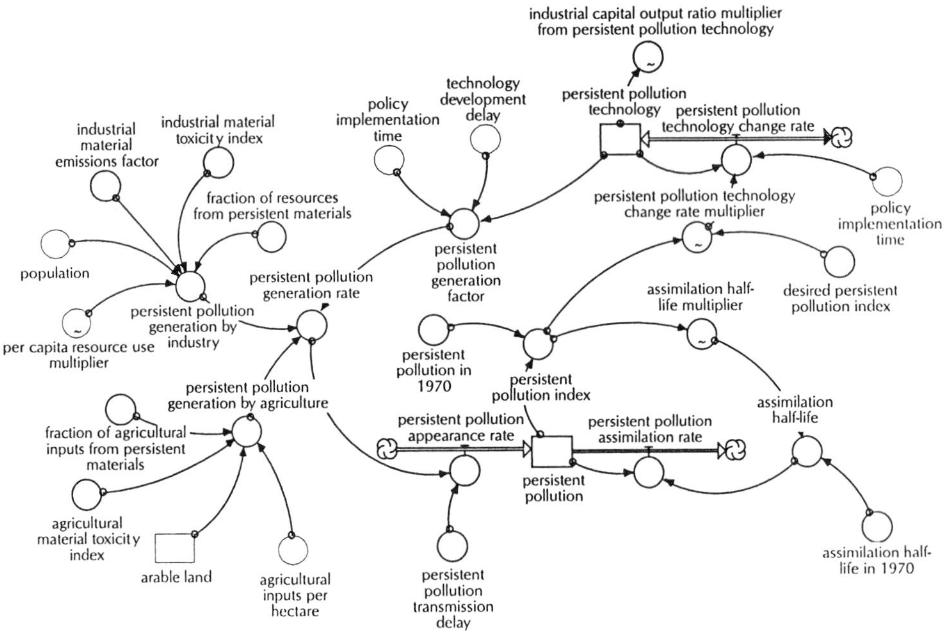

Figure A-2 NONRENEWABLE RESOURCES

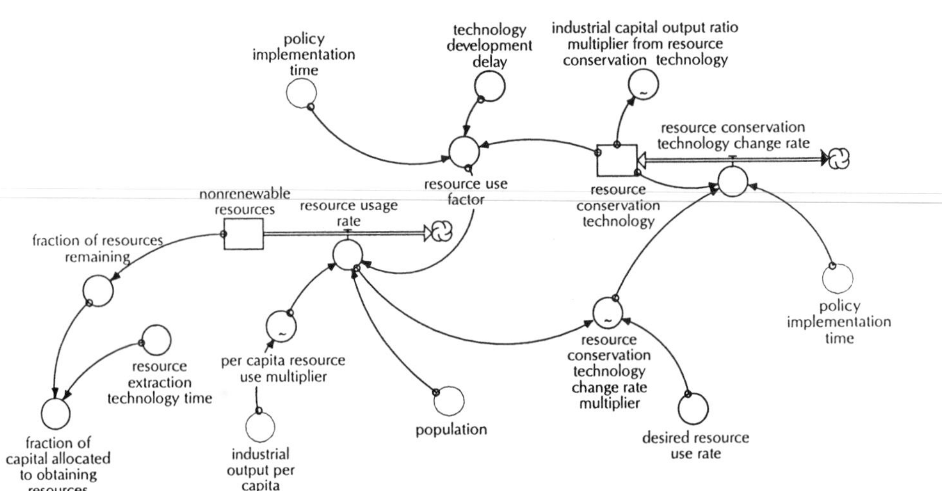

implications of other assumptions, or create computer-based teaching materials for use in helping others to understand principles of systems analysis or to consider global futures. That will require a detailed knowledge of the model.

In this appendix we provide the information you will need to get started. First, we present a set of nine diagrams that portray the major elements and relationships in the model. Second, we describe the equation changes we used to convert the version of World3 employed in *The Limits to Growth* into World3/91, the version of the model that was used in this book. We translated the original DYNAMO equations to STELLA equations, so that our scenarios could be generated on a Macintosh microcomputer. Then we made seven changes in World3's constants and table functions to reflect numerical changes revealed by analysis of the past twenty years of global data. We also altered our representation of the means by which technological change influences the coefficients of the model.

Finally, we describe how you can acquire the three tools you need to experiment with and revise World3/91: the technical documentation for the model, a set of World3/91 computer equations for Macintosh or IBM-compatible computers, and the corresponding simulation software suited to your computer.

The Elements and Relationships in World3

The model is comprised of five sectors: persistent pollution, nonrenewable resources, population, agriculture (food production, land fertility, and land development and loss), and economy (industrial output, services output, and jobs). The principal elements and relationships of these sectors are illustrated in Figures A1 through A9. The precise element names used in the STELLA equations of World 3/91 are not employed in these nine illustrations. To make these diagrams as comprehensible as possible, we have eliminated all the abbreviations and equation numbers that were used in the computer model equations.

We make no attempt in this appendix to explain or justify World3. Except for the eight changes described in the following section, the nu-

Figure A-3a POPULATION SECTOR

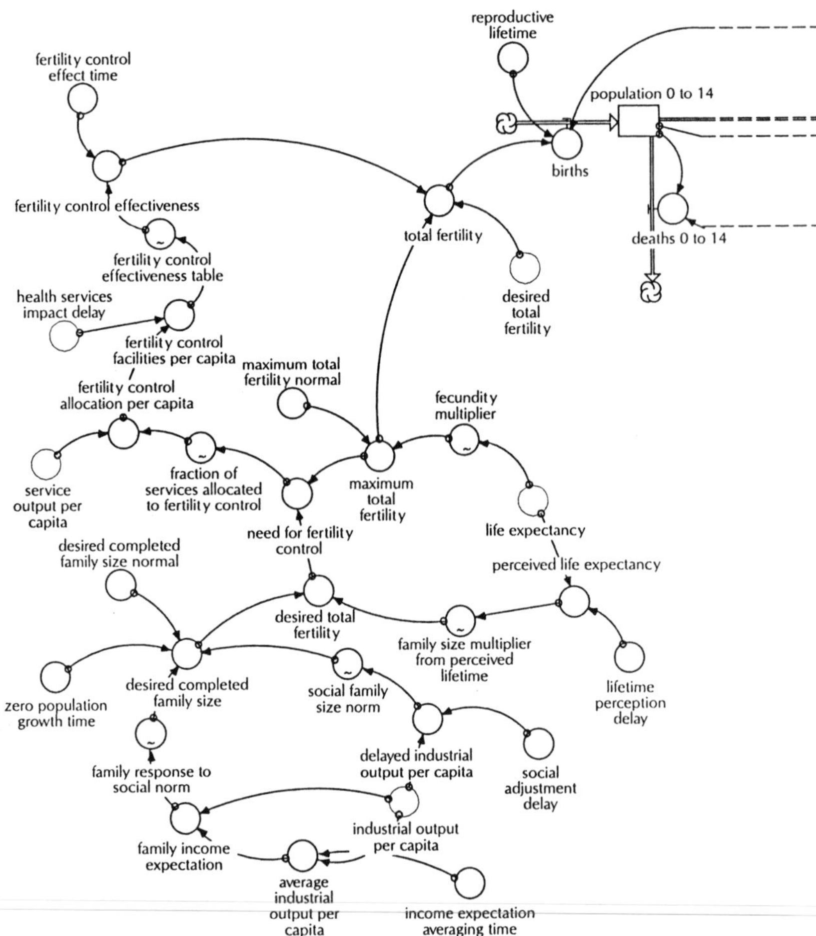

merical values of all coefficients and the precise nature of all relationships in the original version of World3 are documented in our 637-page technical report, *Dynamics of Growth in a Finite World.*[2] That book also includes a full equation listing of the World3 model and the changes necessary to generate all the scenarios presented in *The Limits to Growth.*

Figure A-3b POPULATION SECTOR

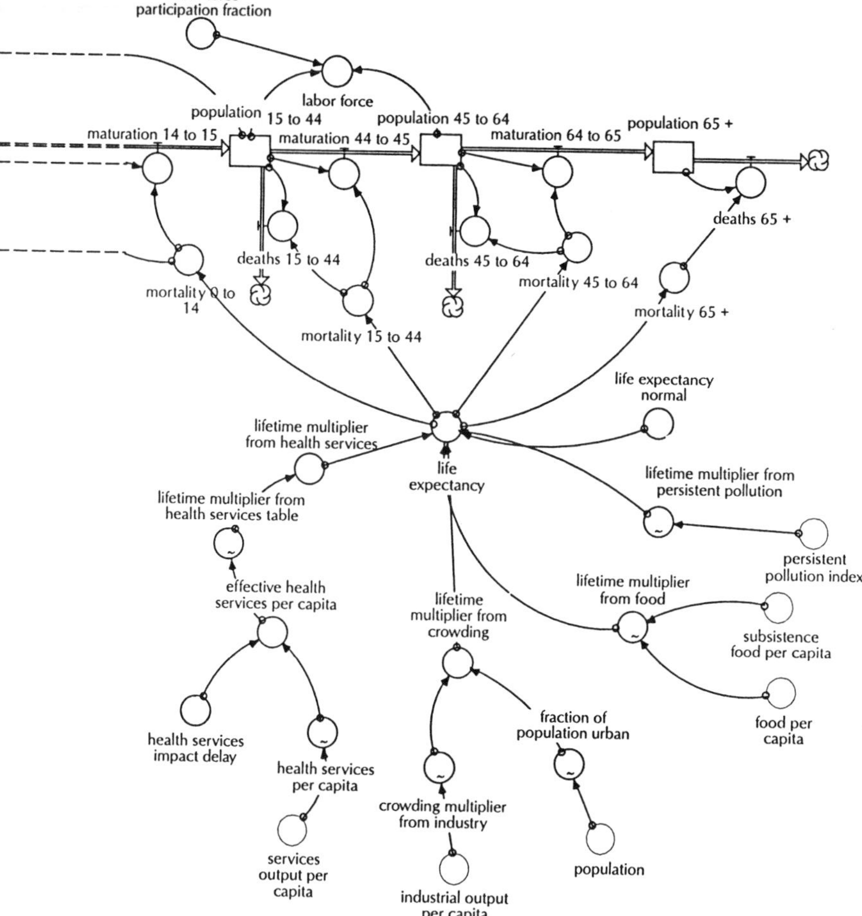

Creating the STELLA Version of World3

The original World3 model was created in the DYNAMO simulation language.[3] There is no DYNAMO compiler for the Macintosh family of microcomputers, so we started our analysis by converting World3 into STELLA.[4] One of STELLA's advantages over DYNAMO is its acceptance of longer and more intelligible variable names. We maintained the structure of World3, but renamed the elements. Because

Figure A-4a FOOD PRODUCTION

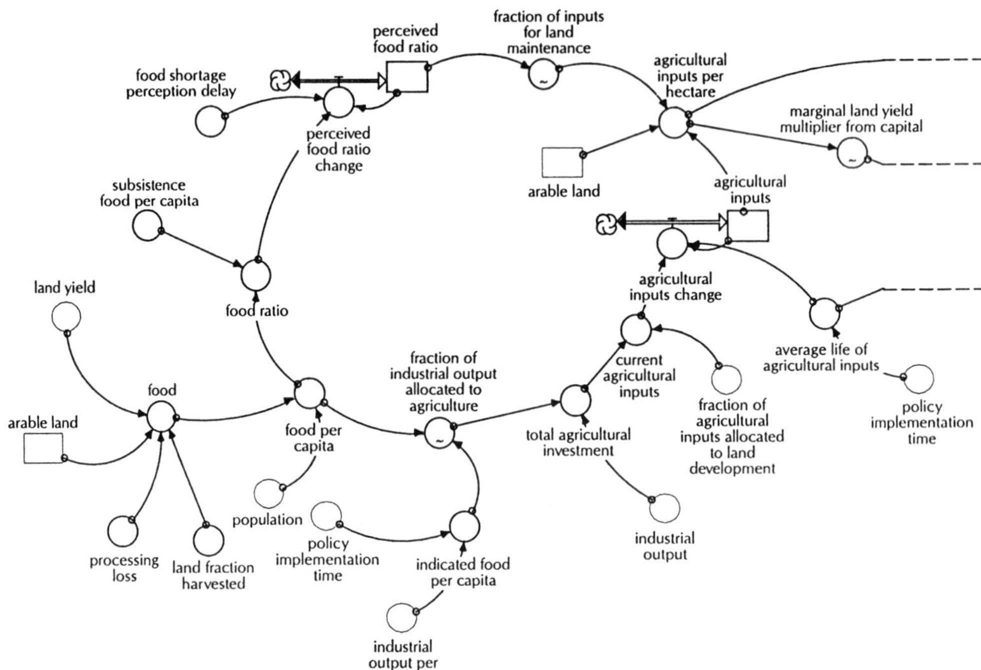

STELLA has a slightly different set of special functions than the one used in DYNAMO, it was necessary to make a few changes in the World3 equations to implement them in STELLA. These equation changes are as follows:

- In four equations, the third-order material delay function (DELAY3) in DYNAMO was replaced with a third-order information delay (SMTH3) in STELLA. This change has no effect so long as the period of the delay remains constant throughout the simulation, which is the case in World3 and World3/91.
- In three cases, the first-order information delay function (SMOOTH in DYNAMO, SMTH1 in STELLA) could not be used in STELLA, because it would have introduced a simultaneous equa-

Figure A-4b FOOD PRODUCTION

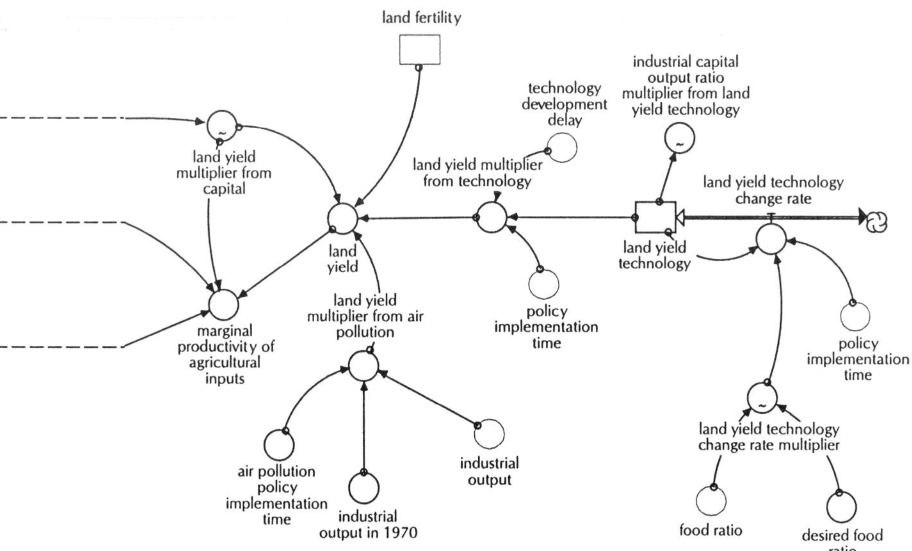

tion loop. It was replaced by an equivalent, explicit stock-flow structure.

- The CLIP function in DYNAMO, which switches between two inputs at a specified time, was replaced with an IF/THEN/ELSE construction in STELLA.

None of these changes in the equations alters the numerical output or behavior of the model. The STELLA version of World3 produces output that is identical to the DYNAMO output found in *The Limits to Growth* and *Dynamics of Growth in a Finite World.*

Changing World3 to Create World3/91

Once World3 had been translated to operate with STELLA, we examined its performance over the period 1970 to 1990. Generally, the behavior of key variables in the model, such as population and food production, was similar to historical data for these variables over the past two decades. We did not find any reason to make structural

Figure A-5 LAND FERTILITY

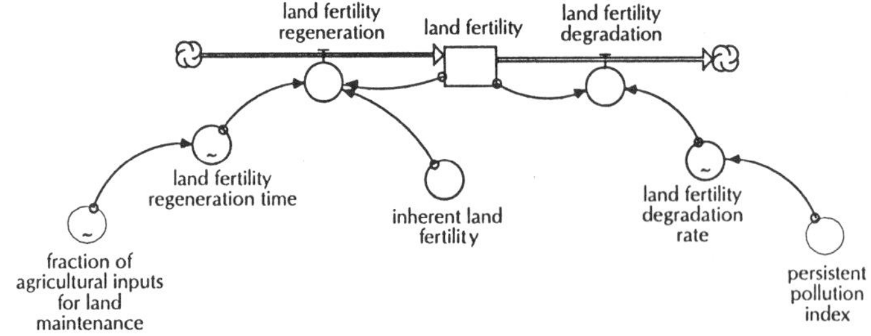

Figure A-6 LAND DEVELOPMENT AND LAND LOSS

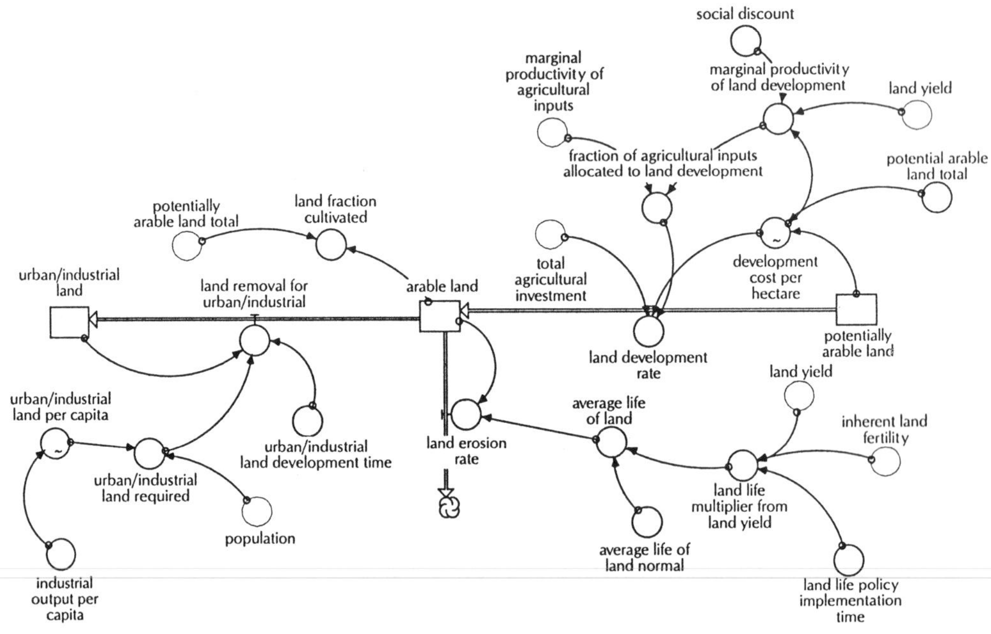

changes in the model equations. However, empirical data on performance of the global system during the past two decades did point to several coefficients and table functions that needed minor revision. As a result we made seven parameter changes.

Figure A-7 INDUSTRIAL OUTPUT

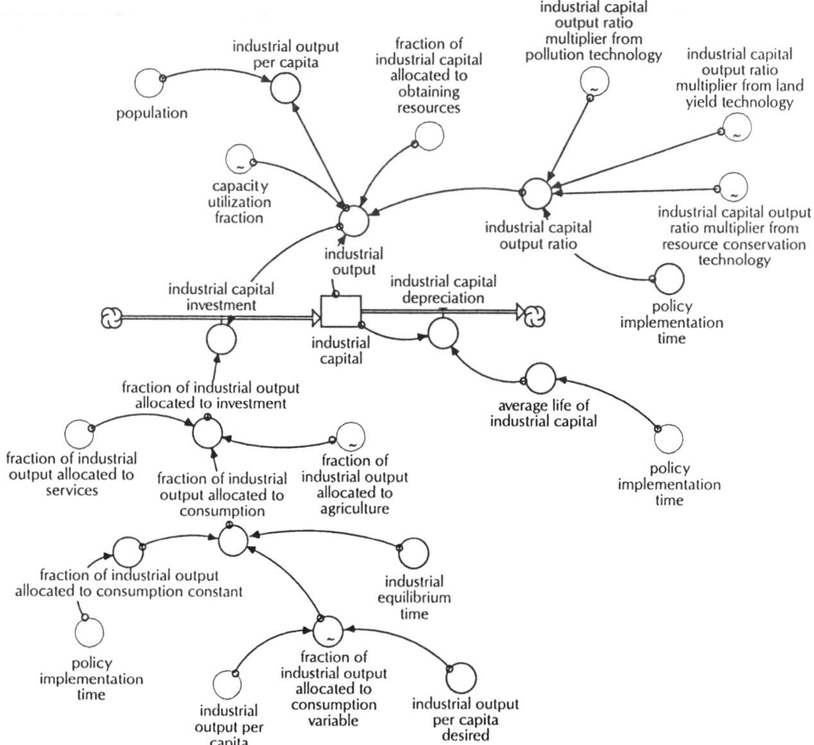

Agriculture

It appears from the past twenty years of data that the coefficients in the original version of World3 underestimated the impact of erosion on the stock of arable land and underestimated the effect of increased agricultural inputs in raising land yields. Because these two errors had opposite effects, the total food output projected by World3 over the period 1970 to 1990 was still in good correspondence with historical data. To make the model parameters more realistic, however, we reduced the normal average life of land in World3/91 from 6000 years to 1000 years, and increased part of the table function defining the impact of agricultural inputs on land yield.

Figure A-8 SERVICES OUTPUT

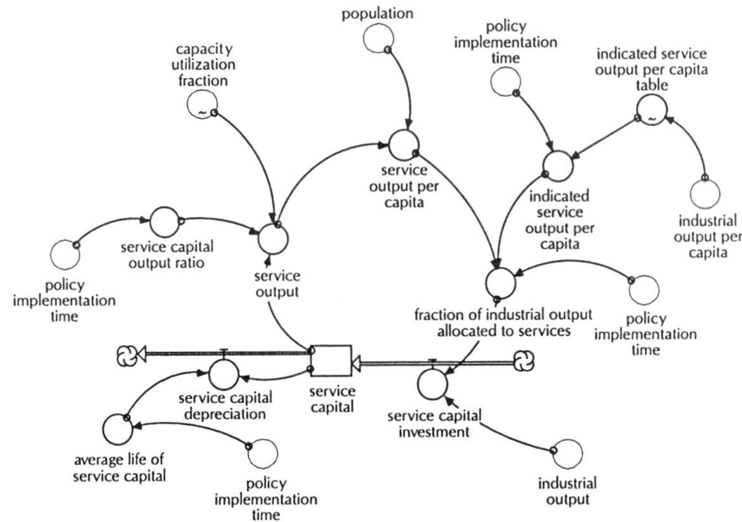

Figure A-9 JOBS

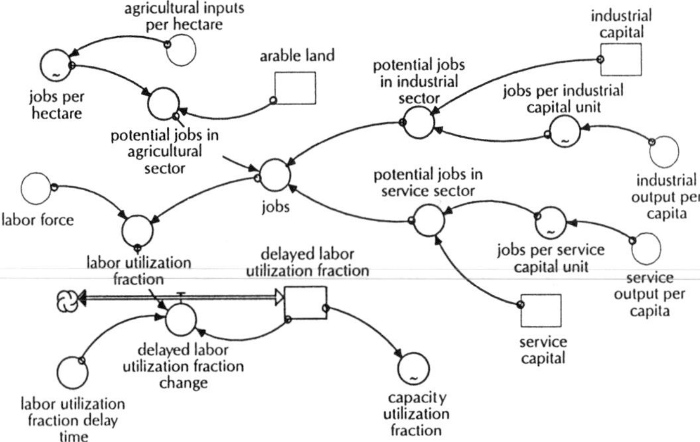

Figure A-10 REVISED TABLE OF RELATION BETWEEN AGRICULTURAL INPUTS AND LAND YIELD

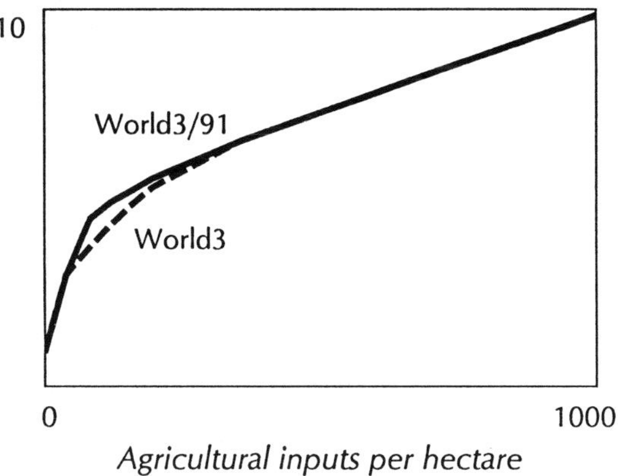

Figure A-11 REVISED TABLE OF RELATION BETWEEN HUMAN HEALTH AND FERTILITY

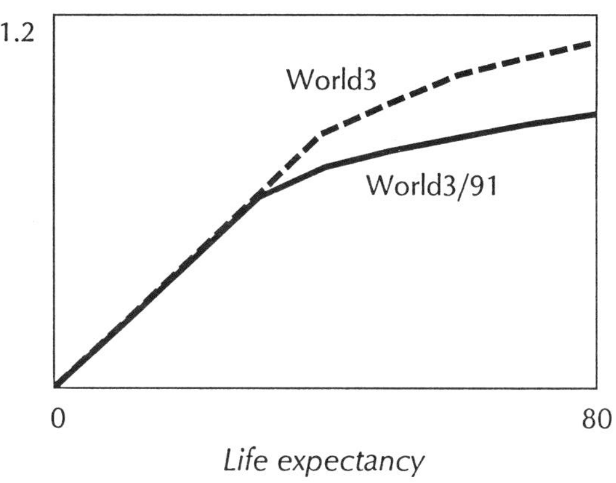

Population

World3 underestimated the rate of decline in both global birth and death rates during the past two decades. Again these offsetting errors left the population projection very close to actual data during the past

Figure A-12 REVISED TABLE OF RELATION BETWEEN FOOD CONSUMPTION AND LIFETIME

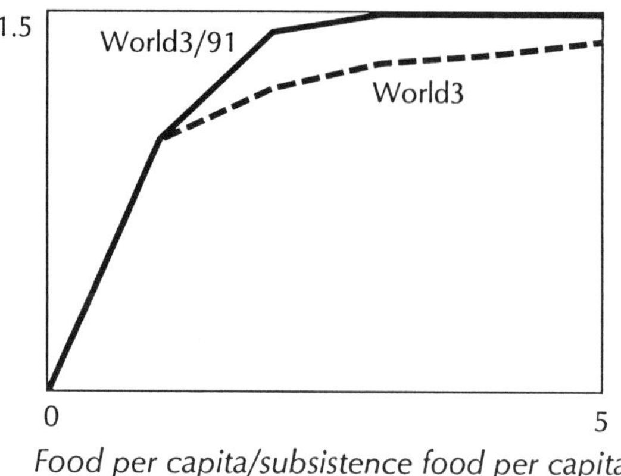

Figure A-13 REVISED TABLE OF RELATION BETWEEN HEALTH SERVICES AND LIFE EXPECTANCY

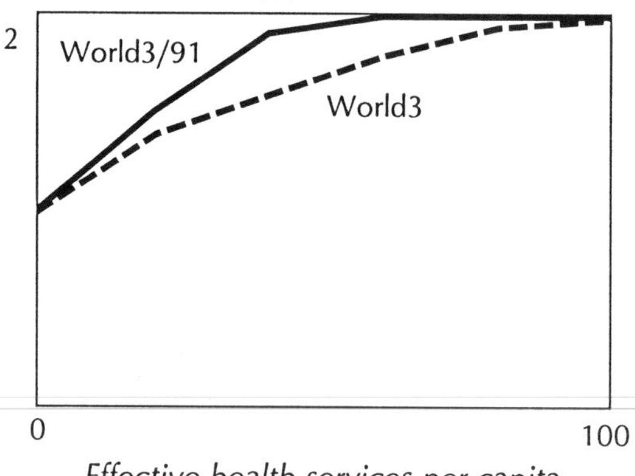

two decades. For World3/91 we revised the model's coefficients to make them more representative of historical experience. We decreased the fecundity multiplier, which determines impact of human health on fertility.

We decreased completed family size normal from 4.0 to 3.8. We revised upward the influence of food consumption on lifetime. We in-

Figure A-14 Revised Table of Relation between Industrial Output and Per Capita Use of Nonrenewable Resources

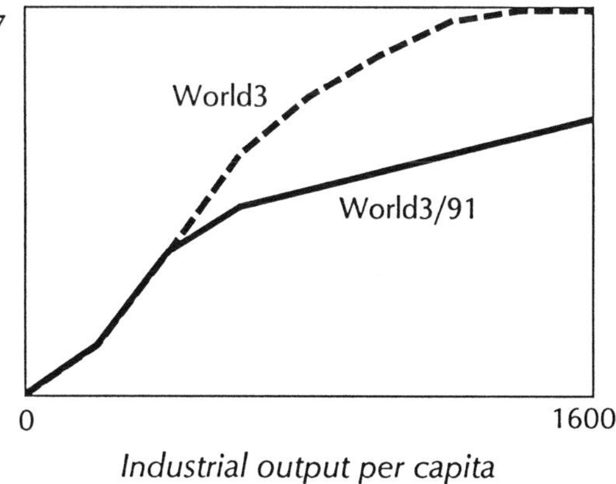

creased the impact of health services in raising the life expectancy, even at low levels of services per person.

Resources

Experience in the most industrialized nations during the past twenty years suggests that our original estimates of resource use per unit of industrial output may have been too high. We reduced the estimates of resource consumption at higher levels of industrial output per capita.

Technology

In *The Limits to Growth*, policies to test the possible implications of new technologies were generally introduced through exogenous and instantaneous shifts from one coefficient to another or from one table function to another. To produce the scenarios in this report we have instead used the adaptive technology structure that was described in Chapter 7 of our technical report. In the adaptive approach there is a system goal, for example, a desired level of persistent pollution. When the actual system state in the model deviates from the goal in a negative direction, for example, persistent pollution becomes greater than desired, capital is allocated to new technologies. After a delay to repre-

Figure A-15 The STELLA Formulation Used to Represent the Adaptive Technology Affecting Land Yield in World3/91

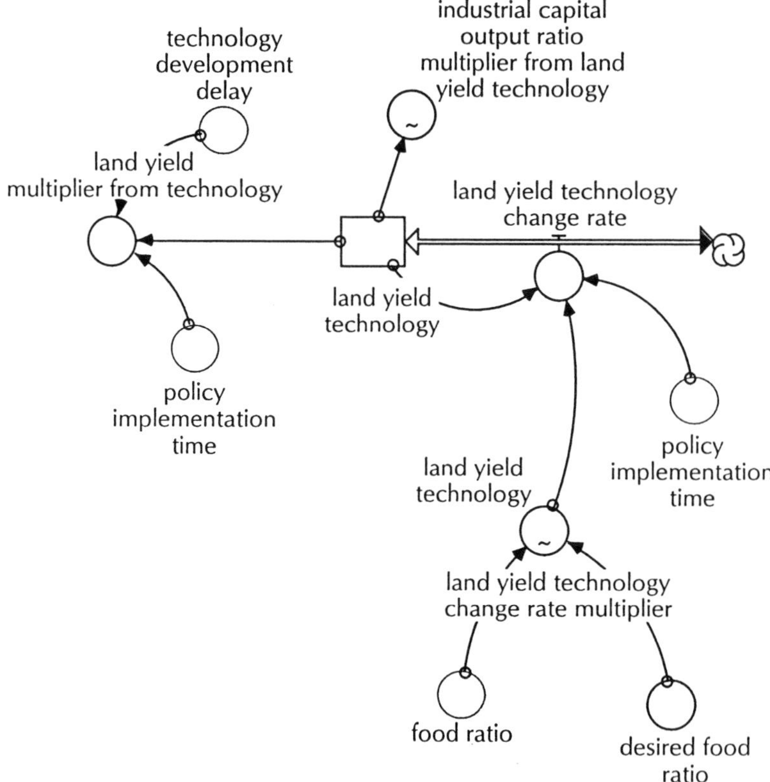

sent the effects of technology development and diffusion, coefficients in the model are adjusted to represent the impact of the new technology, which acts to reduce the problem. This process of adaptive technology development continues until the model world again achieves its goal. To achieve technology advances, modest amounts of industrial capital are diverted from producing industrial output to acquiring and maintaining the new technology. The STELLA flow diagram segment in Figure A-15 indicates how this approach was implemented for technologies to improve land yield.

Table A-1 VARIABLE SCALES IN THE WORLD3/91 SCENARIOS

Variable	*Low value*	*High value*
State of the world		
Population (persons)	0	13×10^9
Total food production (vegetable-equivalent kg/year)	0	6×10^{12}
Total industrial output (1968-$/year)	0	4×10^{12}
Index of persistent pollution (1 = 1970)	0	40
Nonrenewable resources (1 x 10 = supply in 1990)	0	2×10^{12}
Material standard of living		
Food per capita (vegetable-equivalent kg/person/year)	0	1000
Consumer goods per capita (1968-$/person/year)	0	250
Services per capita (1968-$/person/year)	0	1000
Life expectancy (years)	0	90

Scenarios

After the seven coefficient changes were introduced, but with all technology policies set to zero, we had a version of World3/91 that corresponded to the standard run of World3 in *The Limits to Growth.* This was designated Scenario 1 in this report. Scenario 1 was then altered to create all the other numbered scenarios and the Unlimited Technology Scenario.

The World3/91 Scenario Scales

The values of five variables are plotted in the "State of the World" graph for each scenario: four variables are plotted in the "Material Standard of Living" graph. We did not put the numerical scales on the vertical axis of each graph because we do not consider the precise values of the variables in each scenario to be very significant. However, we provide those scales here for readers with a more technical interest in the simulations. The nine variables are plotted on very different scales, but those scales are held constant throughout all fourteen scenarios. To interpret these ranges you will have to study the description of each variable that is presented in the technical report.

The Technical Documentation for World3

Dynamics of Growth in a Finite World is an exhaustive technical description of World3. The book describes the history and the purpose of the model, defines each variable, describes and justifies every causal hypothesis incorporated in World3, gives the detailed listing of equations in the computer language DYNAMO, and provides a large number of simulations to illustrate the behavior of the model's five sectors. The book, written by Dennis L. Meadows and other members of the original *Limits to Growth* team, is distributed by Productivity Press, 2067 Massachusetts Avenue, 4th Floor, P.O. Box 3007, Cambridge, MA 02140. Telephone: (617) 497-5146. Telefax: (617) 868-3524.

The Computer Simulation Software

World3 was originally programmed in DYNAMO, a simulation language developed at the Massachusetts Institute of Technology specifically for analyzing system dynamics models. This is still the language of choice for those working on mainframe computers or on IBM-compatible microcomputers. For this report we translated the model into a form that can be simulated using STELLA.

You can purchase the DYNAMO and STELLA compilers from their respective vendors. DYNAMO is sold by Pugh Roberts, Inc., 41 William Linskey Way, Cambridge, MA 02142. Telephone: (617) 864-8880. Telefax: (617) 661-7418. STELLA is sold by High Performance Systems, Inc., 45 Lyme Road, Suite 300, Hanover, NH 03755. Telephone: (603) 643-9636. Telefax: (603) 643-9502.

A package containing a computer disk with the equations of World3/91 formatted for use with either compiler and a detailed list of the changes required to produce each scenario in this book can be purchased from the Laboratory for Interactive Learning, Hood House, University of New Hampshire, Durham, NH 03824. Telephone: (603) 862-2186. Telefax: (603) 862-1488.

Complementary Teaching Materials

The Laboratory for Interactive Learning has also published numerous teaching materials related to environmental protection and sustainable development. Included are books, educational games, and other materials that directly complement *Beyond the Limits* in teaching. You can request a free brochure that includes information on each item and provides order information.

Endnotes

Preface

1. Donella H. Meadows et al., *The Limits to Growth* (New York: Universe Books, 1972).

2. Yes, there was a World1 and also a World2. World1 was the prototype model first sketched out by MIT Professor Jay Forrester in response to The Club of Rome's inquiry about interconnections among global trends and problems. World2 is Forrester's final documented model, described in Jay W. Forrester, *World Dynamics* (Cambridge: Wright-Allen Press, 1971). Distributed by Productivity Press. The World3 model was developed from World2, primarily by elaborating its structure and extending its quantitative data base. Forrester is the intellectual source of the World3 model and of the system dynamics modeling method it employs.

3. There were also two technical books, Dennis L. Meadows et al., *The Dynamics of Growth in a Finite World* (Cambridge: Wright-Allen Press, 1974) and Dennis L. Meadows and Donella H. Meadows, *Toward Global Equilibrium* (Cambridge: Wright-Allen Press, 1973). Both books distributed by Productivity Press. The first is a complete documentation of the World3 computer model;

the second is a description of several auxiliary studies and submodels made as input to the global model.

4. These headlines are from, respectively, the Saskatoon *Star-Phoenix* (Canada), the Cleveland *Plain Dealer* (USA), and the Tokyo *Mainichi Daily News* (Japan).

5. *Limits*, p. 24.

6. Thomas Vargish, "Why the Person Sitting Next to You Hates *Limits to Growth*," *Technological Forecasting and Social Change* 16 (1980): 179–189.

A Note on Language

1. Robert Goodland, Herman Daly, and Salah El Serafy, introduction to *Environmentally Sustainable Economic Development: Building on Brundtland*, The World Bank Environment Working Paper, no. 46, July 1991, 2–3.

Chapter 1

1. Aurelio Peccei, *One Hundred Pages for the Future* (New York: Pergamon Press, 1981), 15. Peccei was an Italian industrial leader and the founder of The Club of Rome.

2. The commission is also known as the Brundtland Commission, after its leader, Gro Harlem Brundtland, Prime Minister of Norway. World Commission on Environment and Development, *Our Common Future* (Oxford: Oxford University Press, 1987), 8.

Chapter 2

1. Thomas E. Lovejoy, Plenary Address to the Annual Meeting of the American Institute of Biological Sciences, 14 August 1988. Lovejoy is a tropical ecologist and the Assistant Secretary for External Affairs of the Smithsonian Institution.

2. This exercise is described in J. Scott Armstrong, *Long-Range Forecasting* (New York: John Wiley & Sons, 1985), 102.

3. We are indebted to Robert Lattes for telling us this riddle. If you want to experience the sudden-filling-up phenomenon personally, try eating 1 peanut on the first day of the month, 2 peanuts on the second day, 4 peanuts on the third, and so on. Try to guess in advance on what day you'll have to give up this exercise!

4. Population Reference Bureau, *World Population Data Sheet* (Washington, DC, 1991).

5. Lester Brown, *State of the World 1992* (New York: W. W. Norton, 1992), 176.

6. United Nations Food and Agriculture Organization (hereafter cited as FAO), *The State of Food and Agriculture 1990* (Rome: United Nations, 1991), 14.

Chapter 3

1. World Commission on Environment and Development (hereafter cited as WCED), *Our Common Future* (Oxford: Oxford University Press, 1987), 8.
2. Herman Daly, "Toward Some Operational Principles of Sustainable Development," *Ecological Economics* 2 (1990): 1–6.
3. "The Hunger Report 1990" issued annually by the Alan Shawn Feinstein World Hunger Program. (Available from Brown University, Box 1831, Providence, RI 02912.)
4. "The Hunger Report 1991" (see note 3 above).
5. G. M. Higgins et al., *Potential Population Supporting Capacities of Lands in the Developing World* (Rome: FAO, 1982). This technical study is summarized in a nontechnical report by Paul Harrison, *Land, Food, and People*, FAO, Rome, 1984. The factor of 16 is based on extremely optimistic assumptions, and it applies only to developing countries, which are starting from a low-yield base. The FAO has not done a similar study for the lands of the industrialized countries.
6. Food from the sea is even more limited than land-based food, and its use is also close to or beyond limits, as we describe in Chapter 6. Futuristic schemes for non-land-based food—yeast in vats and so forth—will be marginal as major food sources, mainly because of the energy and capital they would demand. Food that is not grown primarily on land and through photosynthesis from the sun's energy would be even more unsustainable than the present food system.
7. World Resources Institute, *World Resources 1990–91* (New York and Oxford: Oxford University Press, 1990), 88.
8. Lester Brown, *State of the World 1991* (New York: W. W. Norton, 1991), 3.
9. WCED, 1987, op. cit., 125.
10. U.N. Population Fund, *The State of World Population 1990* (New York: 1990), 10.
11. See, for example, Michael J. Dover and Lee M. Talbot, *To Feed the Earth: Agro-Ecology for Sustainable Development* (Washington, DC: World Resources Institute, June 1987).
12. The literature of "organic," "low-input," or "ecological" agriculture is enormous. For worldwide examples contact the International Federation of Organic Agricultural Movements, Ökozentrum Imsbach, W-6695, PholeyTheley, Germany. For examples of U.S. commercial organic farms, see any issue of *New Farm*, Rodale Press, Emmaus, PA. For a review of organic farming in the

United States, see U.S. National Research Council, *Alternative Agriculture* (Washington, DC: National Academy Press, 1989).

13. For a summary of problems and potentials, see Pierre R. Crosson and Norman J. Rosenberg, "Strategies for Agriculture," *Scientific American* (September 1989): 128.

14. Of this amount about 2100 cubic kilometers are withdrawn and consumed, primarily in irrigation. The remaining 1400 cubic kilometers are returned to streams, usually in polluted condition. See World Resources Institute, *World Resources 1990-91* (New York and Oxford: Oxford University Press, 1990), 170.

15. The actual capacity of human-made reservoirs is about 6000 cubic kilometers, but only about half of that is actually available as sustainable flow. For a summary of a recent Russian report on global water storage basins, see *World Resources 1990–91*, op. cit., 170.

16. World water withdrawals have been increasing by 4% to 8% per year for decades, but they are beginning to stabilize in the industrial countries. Withdrawals in the less-industrialized countries are expected to go on increasing at 2% to 3% per year.

17. For some current case studies see Malin Falkenmark, "Fresh Waters as a Factor in Strategic Policy and Action," in Arthur H. Westing, ed., *Global Resources and International Conflict* (Oxford: Oxford University Press, 1986).

18. See, for example, Sandra Postel, "Saving Water for Agriculture" in Lester Brown, *State of the World 1990* (New York: W. W. Norton, 1990), 45–47; Jayanta Bandyopadhyay, "Riskful Confusion of Drought and Man-induced Water Scarcity," *Ambio* 18 (1989): 284-292; Danilo Anton, "Thirsty Cities," *IDRC Reports* (October 1990); Chandran Nair, "Bangkok's Deteriorating Groundwater" (Paper presented at 14th WEDC Conference, Kuala Lumpur, 1988).

19. Forest statistics quoted in this section are taken from Sandra Postel and John C. Ryan, "Reforming Forestry" in Lester Brown, *State of the World 1991* (New York: W. W. Norton, 1991), 74–92; and from World Resources Institute, 1990, op. cit., 101–120.

20. Postel and Ryan, op. cit., 75.

21. International Institute of Applied Systems Analysis (hereafter cited as IIASA), *Options* (Laxenburg, Austria: September 1990): 4.

22. IIASA, *Options*, op. cit., 11.

23. Postel and Ryan, op. cit., 88.

24. WCED, 1987, op. cit., 149.

25. See, for example, "Extinction: Are Ecologists Crying Wolf?" *Science* (18 August 1991): 736–and other articles in the same issue, which express the serious concerns of ecologists.

26. Peter M. Vitousek et al., "Human Appropriation of the Products of Photosynthesis," *BioScience* 36 (1986): 368.
27. The Netherlands is said to "occupy" somewhere between 5 and 7 times its own territory, largely because of imports of animal fodder from Third World countries. See Rijksinstituut voor Volksgezondheit en Milieuhygiene (hereafter cited as RIVM), *National Environmental Outlook, 1990–2010* (Bilthoven, the Netherlands: RIVM, 1991).
28. Commercial energy means that sold in a market; it does not count the energy used by people who gather wood, dung, and other biomass for their own use. "Noncommercial" energy sources are mostly renewable, though they may not be harvested sustainably. They account for approximately 15% of the energy consumption of the human population.
29. Ged R. Davis, "Energy for Planet Earth," *Scientific American* (September 1990): 55.
30. Quoted in Christopher Flavin and Nicholas Lenssen, "Designing a Sustainable Energy System" in Lester Brown, *State of the World 1991* (New York: W. W. Norton, 1991), 21.
31. "Production" is a misleading word for the process of taking fossil fuels out of the ground. Nature is the producer of these fuels, over millions of years. Human beings do not "produce" them; they extract, exploit, harvest, pump, mine, or take them. However, production is the word commonly used, especially in such terms as "reserve/production ratio," so we have used it as well.
32. Of course the capital plants for discovery, mining, pumping, transporting, and refining also burn fuels. If there were no other limits, the ultimate limit to the use of fossil fuels would come at the point where it takes as much energy to get them as they contain. See Charles A. S. Hall and Cutler J. Cleveland, "Petroleum Drilling and Production in the United States: Yield per Effort and Net Energy Analysis," *Science* (6 February 1981): 576.
33. For gas reserve see H. J. M. de Vries, "The Carbon Dioxide Substitution Potential of Methane and Uranium Reserves" in P. Okken, R. Swart, and S. Zwerver, eds., *Climate and Energy: The Feasibility of Controlling CO_2 Emissions* (Dordrecht: Kluwer Academic Publishers, 1989). For patterns of reserve overestimation see John D. Sterman, George P. Richardson, and Pål Davidsen, "Modeling the Estimation of Petroleum Resources in the United States," *Technological Forecasting and Social Change* 33 (1988): 219–248.
34. This information, and most of the data we quote on this topic, comes from Amory Lovins and the Rocky Mountain Institute. For more detailed information on energy efficiency options in transportation, industry, and buildings, see *Scientific American* (September 1990).

35. For varying estimates, see Jose Goldemberg et al., *Energy for a Sustainable World* (Washington, DC: World Resources Institute, 1987); *Energy 2000: A Plan of Action for Sustainable Development* (Danish Ministry of Energy, April 1990); Amulya K. N. Reddy, "Energy Strategies for a Sustainable Development in India" (Paper presented at the conference Global Collaboration on a Sustainable Energy Development, Copenhagen, April 1991); Arnold P. Fickett, Clark W. Gellings, and Amory Lovins, "Efficient Use of Electricity," *Scientific American* (September 1990): 64; M. Grubb et al., *Energy Policies and the Greenhouse Effect, Volume Two: Country Studies and Technical Options*, The Royal Institute of International Affairs (Worcester: Billings & Sons, Ltd., 1991).
36. Rocky Mountain Institute newsletter, Spring 1991. Both figures are in 1987 dollars.
37. For a well-annotated review, see Christopher Flavin and Nicholas Lensson, "Designing a Sustainable Energy System" in Lester Brown, *State of the World 1991* (New York: W. W. Norton, 1991).
38. Meridian Corporation, "Characterization of U.S. Energy Resources and Reserves," prepared for Deputy Assistant Secretary for Renewable Energy, DOE, Alexandria, VA, June 1989; Idaho National Engineering Laboratory et al., "The Potential of Renewable Energy: An Interlaboratory White Paper, prepared for the Office of Policy, Planning and Analysis, DOE, in support of the National Energy Strategy," Solar Energy Research Institute, 1990; DOE, EIA, *Annual Energy Review 1989* (Washington, DC, 1990).
39. The most promising storage mechanism may be hydrogen, made from solar-electric splitting of water molecules. Hydrogen may also be the answer to vehicle propulsion in the future. For a review, see Joan M. Ogden and Robert H. Williams, *Solar Hydrogen* (Washington, DC: World Resources Institute, October 1989).
40. Amory B. Lovins, *Openpit Mining* (London: Earth Island, 1973), 1.
41. For a systematic examination of these possibilities, see John E. Tilton, ed., *World Metal Demand* (Washington, DC: Resources for the Future, 1990).
42. For example, the U.S. economy in 1985 generated 187 million metric tons municipal solid waste, 628 million tons industrial waste (265 million of which were hazardous), 72 million tons energy waste, 1400 million tons agricultural waste, 1300 million tons mining waste (excluding coal), 98 million tons demolition waste, 8.4 million tons sewage sludge. See OECD, *Environmental Data, Compendium 1989* (Paris, 1989), 155.
43. Nanotechnology is the design and use of molecules as machines, the assembly of products in a controlled way molecule by molecule–which is the way living things are assembled. Biotechnology is a particularly sophisticated subset

of nanotechnology that concentrates on the design and use of DNA molecules. For an enthusiastic description of the possibilities of nanotechnology, see K. Eric Drexler and Chris Peterson, *Unbounding the Future: The Nanotechnology Revolution* (New York: William Morrow, 1991). See also "Materials for Economic Growth," *Scientific American* (September 1986) and the special issue of *Science* (29 November 1991).

44. Earl Cook, "Limits to Exploitation of Nonrenewable Resources," *Science* (20 February 1976).

45. The United States, Japan, Great Britain, France, Germany, Italy, and Canada.

46. Barry Commoner, *Making Peace with the Planet* (New York: Pantheon Books, 1990).

47. California Air Resources Board, *Air Review* 3 (1991): 45.

48. I. F. Langeweg, *Concerns for Tomorrow* (Bilthoven, the Netherlands: RIVM, 1989).

49. *National Environmental Outlook 1990–2010* (Bilthoven, the Netherlands: RIVM, 1991).

50. WCED, 224.

51. WCED, 226.

52. *Nordic Environment* 9 (September 1991): 2.

53. See William K. Stevens, "Northern Hemisphere Snow Cover Found to Be Shrinking," *New York Times*, 30 October 1990, C4; "The Ghosts of Coral Past," *U.S. News and World Report*, 23 September 1991; Keith Schneider, "Ranges of Animals and Plants Head North," *New York Times*, 13 August 1991, C1; *Science* (12 October 1990): 213.

54. For a careful compendium of world scientific opinion on global climate change assembled by several hundred working scientists from 25 countries, see World Meterological Organization/United Nations Environment Programme Intergovernmental Panel on Climate Change, *Climate Change: The IPCC Scientific Assessment* (Cambridge: Cambridge University Press, 1990).

55. These data come from ice cores drilled deep into the Antarctic ice sheet. The polar ice has accumulated over thousands of years, layer after layer, and in each layer are trapped tiny air bubbles, preserved from prehistoric time. Isotopic analysis can date the core layers and provide clues to past temperatures; direct analyses of the air bubbles give the carbon dioxide and methane concentrations.

56. Ken Geiser, "The Greening of Industry," *Technology Review* (August/ September 1991): 64.

57. Geiser, op. cit., 64.

58. For an extended discussion of the IPAT formula, see Paul R. Ehrlich and Anne H. Ehrlich, *Healing the Planet* (Reading, MA: Addison-Wesley, 1991).
59. We have adapted this formulation from one originally put forward by Amory Lovins. The estimates for time-scale and scope for long-term change are his.
60. Lester Thurow, *Technology Review* (August/September 1986). Thurow is the Dean of MIT's Sloan School of Management.

Chapter 4

1. William R. Catton, Jr., *Overshoot: The Ecological Basis of Revolutionary Change* (Urbana: University of Illinois Press, 1980).
2. The carrying capacity is the size of population that can be sustained by the environment indefinitely. The concept of carrying capacity was originally defined for relatively simple population/resource systems, such as the number of cattle or sheep that could be maintained on a defined piece of grazing land without degrading the land. For human populations the term "carrying capacity" is much more complex because of the many kinds of resources people take from the environment, the many kinds of wastes they return, and the great variability in technology, institutions, and lifestyles. Carrying capacity is a dynamic concept. A carrying capacity is not constant; it is always changing with weather and other external changes and with the pressure exerted by the species being carried.
3. See, for example, R. Boyd, "World Dynamics, a Note," *Science* (11 August 1972); T. W. Oerlemans et al., "World Dynamics: Social Feedback May Give Hope for the Future," *Nature* 238 (4 August 1972); H. S. D. Cole et al., *Models of Doom* (New York: Universe Books, 1973).
4. Sören Jensen, *New Scientist* 32 (1966): 612.
5. Environment Canada, Department of Fisheries and Oceans, "Toxic Chemicals in the Great Lakes and Associated Effects, Vol. I. Contaminant Levels and Trends," March 1991.
6. E. Dewailly et al., "High Levels of PCBs in Breast Milk of Inuit Women from Arctic Quebec," *Bulletin of Environmental Contamination and Toxicology* 43 (1989): 641–646.
7. P. J. H. Reijnders, "Reproductive Failure in Common Seals Feeding on Fish from Polluted Coastal Waters," *Nature* 324 (4 December 1986): 456.
8. J. M. Marquenie and P. J. H. Reijnders, "Global Impact of PCBs with Special Reference to the Arctic" (Proceedings of the 8th International Congress of Comite Arctique Internationale, Oslo, 18–22 September, 1989, NILU, Lillestrom, Norway).

9. WCED, *Our Common Future (*Oxford: Oxford University Press, 1987), 102.
10. See "New Cause of Concern on Global Warming," *New York Times,* 12 February 1991, C6.
11. W. M. Stigliani, "Chemical Time Bombs," *Options* (Laxenburg, Austria: IIASA, September 1991): 9.
12. Industrial output per capita is the major indicator of material standard of living in the model. A value of 500 1968 dollars per person per year corresponds *very* roughly to the world average in 1990. The variable "consumer goods per capita" used in the plots represents the fraction of undustrial output which is consumer goods, often around 40% of the total. Thus a value of 200 1968 dollars per person per year for consumer goods corresponds roughly to the 1990 world average. Finally, one should remember that the units of the industrial output per capita that are calculated in the model are expressed in 1968 dollars, which were the most recent numbers available when we first developed the equations.

Chapter 5

1. F. Sherwood Rowland, quoted by Paul Brodeur, "Annals of Chemistry: In the Face of Doubt," *New Yorker,* 9 June 1986, 80. Rowland is an atmospheric chemist at the University of California Irvine. He was one of the discoverers of the mechanism by which the ozone layer is eroded.
2. Arjun Makhijani, Annie Makhijani, and Amanda Bickel, *Saving Our Skins: Technical Potential and Policies for the Elimination of Ozone-Depleting Chlorine Compounds* (Washington, DC: Environmental Policy Institute and the Institute for Energy and Environmental Research, September 1988), 83. (Available from The Environmental Policy Institute, 218 O St. SE, Washington, DC 20003.)
3. *Saving Our Skins,* op. cit., 77.
4. See, for example, Robin Russell Jones, "Ozone Depletion and Cancer Risk," *The Lancet* (22 August 1987): 443; "Skin Cancer in Australia," *The Medical Journal of Australia* (1 May 1989); Alan Atwood, "The Great Cover-up," *Time* (Australia) (27 February 1989); Medwin M. Mintzis, "Skin Cancer: The Price for a Depleted Ozone Layer," *EPA Journal* (December 1986).
5. Office of Air and Radiation, U.S. Environmental Protection Agency, *Assessing the Risks of Trace Gases in the Earth's Atmosphere,* vol. VIII (Washington, DC: Government Printing Office, December 1987).
6. Richard S. Stolarski and Ralph J. Cicerone, "Stratospheric Chlorine: A Possible Sink for Ozone," *Canadian Journal of Chemistry* 52 (1974): 1610.

7. Mario J. Molina and F. Sherwood Rowland, "Stratospheric Sink for Chlorofluoromethanes: Chlorine Atomic Catalysed Destruction of Ozone," *Nature* 249 (1974): 810.
8. Quoted in Richard E. Benedick, *Ozone Diplomacy* (Cambridge: Harvard University Press, 1991): 12.
9. J. C. Farman, B. G. Gardiner, and J. D. Shanklin, "Large Losses of Total Ozone in Antarctica Reveal Seasonal ClO/NO_2 Interaction," *Nature* 315 (1985): 207.
10. The period during which scientists were seeing low ozone readings and yet not "seeing" them is described well in Paul Brodeur, op. cit., 1986, 71.
11. J. G. Anderson, W. H. Brune, and M. J. Proffitt, "Ozone Destruction by Chlorine Radicals within the Antarctic Vortex: The Spatial and Temporal Evolution of ClO-O_3 Anticorrelation Based on in Situ ER-2 Data," *Journal of Geophysical Research* 94 (30 August 1989): 11, 474.
12. Mario J. Molina, "The Antarctic Ozone Hole," *Oceanus* 31 (Summer 1988).
13. Du Pont dropped its search for CFC substitutes upon the election of Ronald Reagan as president in 1980.
14. The political process is described clearly and fully by Richard Benedick, who was the chief negotiator for the United States, in *Ozone Diplomacy*, op. cit., 1990.
15. Mario J. Molina, "Stratospheric Ozone: Current Concerns" (Paper presented at the Symposium on Global Environmental Chemistry–Challenges and Initiatives, 198th National Meeting of the American Chemical Society, September 10–15, 1989, Miami Beach, Florida). Molina was the author, with F. Sherwood Rowland, of one of the first scientific papers predicting ozone depletion.
16. The Industrial Coalition for Ozone Layer Protection, 1440 New York Avenue NW, Suite 300, Washington, DC 20005.
17 William K. Stevens, "Summertime Harm to Shield of Ozone Detected over U.S.," *New York Times*, 23 October 1991, 1.

Chapter 6

1. Stewart L. Udall in William R. Catton, Jr., *Overshoot: The Ecological Basis of Revolutionary Change* (Urbana: University of Illinois Press, 1980), xv. Udall was a U.S. Congressman and Secretary of the Interior.
2. Jeremy Bray, *Environment* 14 (May 1972): 44.
3. Julian Simon and Herman Kahn, *The Resourceful Earth* (Oxford: Basil Blackwell, Ltd., 1984), 3.
4. One small reminder to us of the wonderful power of technical advance is that we are writing this book, running the World3 model, preparing graphs

and charts, and laying out pages on desktop and laptop computers. Twenty years ago we wrote *The Limits to Growth* on electric typewriters, we drew graphs by hand, and we needed a huge mainframe computer to run World3.

5. For a particularly thoughtful exploration of technology see C. S. Lewis, "The Abolition of Man," in Herman Daly, *Toward a Steady-State Economy* (San Francisco: Freeman Press, 1973): 321.

6. That assumption was made in 1970, and then we implemented those technologies in the simulated year 1975. By the real year 1990 some of them had begun to be incorporated structurally into the world economy. Therefore we have made some permanent adjustments to the numbers within World3—we have, for instance, significantly reduced resource use per unit of industrial output. These numerical changes are explained in the Appendix. They are are in effect in all model runs shown in this book. Still further reductions in resource use and pollution generation are then implemented in the model through the "turn-on" technologies described above and illustrated in this chapter.

7. Dianne Dumanoski, "Study by the Northeast Fisheries Center Warns That the Level of Cod on the Georges Bank Has Dropped and Is Heading for a Major Collapse," *Boston Globe*, 5 November 1988, 1.

8. For fisheries production see FAO, *Yearbook of Fisheries Statistics: Catches and Landings* (Rome: United Nations, published annually). For estimates of sustainable harvests see M. A. Robinson, *Trends and Prospects in World Fisheries* (Rome: FAO, 1984).

9. *New York Times*, 16 July 1991, C4.

10. The bluefin and shrimp figures come from *Audubon* (September/October 1991): 34, 44.

11. John Kurien and T. R. Thankapan Achari, "Overfishing Along Kerala Coast: Causes and Consequences," *Economic and Political Weekly* (1–8 September 1990).

12. J. F. Caddy and R. C. Griffiths, "Recent Trends in the Fisheries and Environment in the General Fisheries Council for the Mediterranean Area"; FAO, *Studies and Reviews: General Fisheries Council for the Mediterranean* 63 (Rome: United Nations, 1990).

13. Lawrence Ingrassia, "Dead in the Water: Overfishing Threatens to Wipe Out Species and Crush Industry," *Wall Street Journal*, 16 July 1991, 1.

14. The classic analysis of this phenomenon is Garrett Hardin's "The Tragedy of the Commons," *Science* (3 December 1968): 1243.

15. Carl Safina, National Audubon Society, personal communication.

16. Paul Lunven, "The Role of Fish in Human Nutrition," *Food and Nutrition* 8 (1982): 9–18.

17. Paul Ehrlich in R. J. Hoage, ed., *Animal Extinction: What Everyone Should Know* (Washington, DC: Smithsonian Institution Press, 1985), 163.
18. *Nordic Environment* 9 (September 1991).

Chapter 7

1. Herman E. Daly, "Toward a Stationary-State Economy," in John Harte and Robert Socolow, *Patient Earth* (New York: Holt, Rinehart and Winston, 1971) 237.
2. WCED, *Our Common Future*, op. cit.
3. Herman Daly is one of the few people who have begun to think through what kinds of social institutions might work to maintain a desirable sustainable state. He comes up with a thought-provoking mixture of market and regulatory devices. See, for example, Herman Daly, "Institutions for a Steady-State Economy" in *Steady State Economics* (Washington, DC: Island Press, 1991).
4. Aurelio Peccei, *The Human Quality* (Oxford: Pergamon Press, 1977), 85.
5. John Stuart Mill, *Principles of Political Economy*, first published 1848.
6. For an example of "sustainable accounting," see Raul Solorzano et al., *Accounts Overdue: Natural Resource Depreciation in Costa Rica* (Washington, DC: World Resources Institute, December 1991).
7. Lewis Mumford, *The Condition of Man* (New York: Harcourt Brace Jovanovich, 1944), 398–99.

Chapter 8

1. William D. Ruckelshaus, "Toward a Sustainable World," *Scientific American* (September 1989): 167. Ruckelshaus was the administrator of the U.S. Environmental Protection Agency under President Richard Nixon, and he returned to the same position under President Ronald Reagan. He was a member of the World Commission on Environment and Development. Currently he is the chief executive officer of Browning Ferris Industries, Inc.
2. This sense of millennial revolution is in the air. The Ruckelshaus quote that begins this chapter is only one example of the revolutionary talk that is beginning to enter the public discourse. See also the WGBH-TV television series "Race to Save the Planet," Alexander King and Bertrand Schneider, *The First Global Revolution* (New York: Pantheon Books, 1991), and Lester Brown, *State of the World 1992* (New York: W. W. Norton, 1992).
3. Donald Worster, ed., *The Ends of the Earth* (Cambridge: Cambridge University Press, 1988), 11–12.
4. Ralph Waldo Emerson, "War" (Lecture delivered in Boston, March 1838). Re-

printed in *Emerson's Complete Works,* vol. XI (Boston: Houghton, Mifflin & Co., 1887), 177.

5. For a description of how a world with these features could actually develop, see Dennis L. Meadows, ed., *Alternative to Growth-I* (Cambridge: Ballinger, 1977). Distributed by Heronbrook Publications, P.O. Box 844, Durham, NH 03824.

6. For an exploration of such an economics, see Herman Daly and John Cobb, *For the Common Good* (Boston: Beacon Press, 1989).

7. R. Buckminster Fuller, *Critical Path* (New York: St. Martin's Press, 1981).

8. Abraham Maslow, *The Farthest Reaches of Human Nature* (New York: Viking Press, 1971).

9. J. M. Keynes, foreword to *Essays in Persuasion* (New York: Harcourt, Brace and Company, 1932).

10. Aurelio Peccei, *One Hundred Pages for the Future* (New York: Pergamon Press, 1981), 184–85.

Appendix

1. Etienne van de Walle, "Foundations of the Model of Doom," *Science* (26 September 1975): 1077–1078.

2. Dennis L. Meadows et al., *Dynamics of Growth in a Finite World* (Cambridge: Wright-Allen Press, 1974). Distributed by Productivity Press.

3. Alexander L. Pugh, II, *DYNAMO User's Manual,* 5th ed. (Cambridge: MIT Press, 1976). George P. Richardson and Alexander L. Pugh, III, *Introduction to System Dynamics Modeling with DYNAMO* (Cambridge: MIT Press, 1981). Both books distributed by Productivity Press.

4. Barry Richmond et al., *STELLA for Business* (Hanover, NH: High Performance Systems, 1987).

Annotated Bibliography

On the System Dynamics Method and Computer Simulation Programs

Bossel, Hartmut. *Systemdynamik: Grundwissen, Methoden, und BASIC-Programme zur Simulation Dynamischer Systeme.* Braunschweig: Friedrich Vieweg & Sohn, 1987. An introductory text in system dynamics modeling using the BASIC programming language.

—— *Umweltdynamik.* Munich: te-wi Verlag Gmbh, 1985. Applications of system dynamics to environmental and resource systems.

Forrester, Jay W. *Principles of Systems.* Cambridge: Wright-Allen Press, 1968. Distributed by Productivity Press, P.O. Box 30007, Cambridge, MA 02140 (tel: 800-274-9911). The original textbook of system dynamics.

—— *Industrial Dynamics.* Cambridge: Wright-Allen Press, 1961. Distributed by Productivity Press. The founding book of the field, and still a classic, laying out basic systems methods and principles with examples from industrial systems.

Randers, Jørgen, ed. *Elements of the System Dynamics Method.* Cambridge: MIT Press, 1980. Distributed by Productivity Press. A collection of papers by system dynamics practitioners, which spell out the philosophy and paradigms of the field, its principle methods, and how it handles such difficult modeling issues as validation and implementation.

Richardson, George P., and Alexander L. Pugh, III. *Introduction to System Dynamics Modeling with DYNAMO.* Cambridge: MIT Press, 1981. Distributed by Productivity Press. A textbook that introduces the basic principles and techniques of the system dynamics approach. It is especially strong in its discussion of practical skills for conceptualizing, debugging, and testing models.

Roberts, Nancy et al. *Introduction to Computer Simulation, a System Dynamics Modeling Approach.* Reading, MA: Addison-Wesley, 1983. An introductory text with clear language and many illustrative examples and student exercises. The book emphasizes the theory and the graphical and numerical representation of feedback relationships.

Senge, Peter. *The Fifth Discipline: The Art and Practice of the Learning Organization.* New York: Doubleday, 1990. Applications of systems thinking (and other kinds of mind-extending thinking) to the field of business management.

The System Dynamics Review, available from the System Dynamics Society, MIT E40-294, Cambridge, MA 02139. The professional journal of the field of system dynamics.

On Large-Scale Social System Modeling

Barney, Gerald O. et al. *Managing a Nation, The Microcomputer Software Catalog,* second edition. Boulder, CO: Westview Press, 1991. The definitive sourcebook for anyone wishing to start on a program of teaching or research that involves global or national-level models. The text provides comprehensive summaries of available microcomputer programs that help in the representation or simulation of important sectors of a national economy. Chapters on models, games, and simulations related to Agriculture, Demography, Economy, and seven other subject areas start the book. There are also chapters on multisectoral national models and several global models. Finally, a section on data, publications, modeling languages, and principles of modeling gives both beginning and advanced modelers access to a wealth of tools.

Forrester, Jay W. *World Dynamics.* Cambridge: Wright-Allen Press, 1971. Distributed by Productivity Press. A superbly documented description of the World2 model, the predecessor of World3, written by the founder of the system dynamics method.

Meadows, Dennis L. et al. *Dynamics of Growth in a Finite World.* Cambridge: Wright-Allen Press, 1974. Distributed by Productivity Press. A detailed technical description of the World3 computer model. The purpose, underlying paradigms, empirical data, and DYNAMO equations involved in constructing the model are laid out in detail.

—— and Donella H. Meadows. *Toward Global Equilibrium, Collected Papers.* Cambridge: Wright-Allen Press,1973. Distributed by Productivity Press. A collection of thirteen papers written by members of the System Dynamics

Group of the MIT Sloan School of Management to explore the nature and implications of physical growth on a finite planet. Eight of the chapters present system dynamics models in DYNAMO that were created to explore issues of pollution, resource depletion, and population growth in the preparation of World3.

Meadows, Donella H., John Richardson, and Gerhart Bruckmann. *Groping in the Dark, The First Decade of Global Modeling.* New York: John Wiley & Sons, 1981. This book emerged out of a series of global modeling conferences organized by the International Institute for Applied Systems Analysis in Laxenburg, Austria, between April 1974 and September 1977. It compares seven of the principal global models, starting with World3, and explores what is known and what is knowable about the long-term future of the global system through research with this kind of model.

—— and Jennifer M. Robinson. *The Electronic Oracle, Computer Models and Social Decisions.* New York: John Wiley & Sons, 1985. Descriptions of nine important socioeconomic models form the core of this book. The principal modeling methods are described: system dynamics, econometrics, input-output analysis, and linear programming, and there are three chapters that examine the strengths and limitations of formal models for policy analysis. The book concludes with two chapters that address means for enhancing the quality, consistency, objectivity, and utility of these large models.

On Statistical Information about the State of the World

The following publications are issued every year or every other year. We subscribe to them all (and more) and keep them close at hand for reference.

Brown, Lester et al. *State of the World.* New York: W. W. Norton, published annually. A readable and well researched report that focuses each year on a cluster of issues related to sustainability. Agriculture, energy, waste management, population, transportation, and water are frequent themes.

Population Reference Bureau, "World Population Data Sheet," available from PRM, 777 Fourteenth Street NW, Suite 800, Washington, DC 20005. A wall chart that summarizes each year the most recent demographic data (population, birth and death rates, infant mortality, etc.) for the countries of the world.

Sivard, Ruth Leger. *World Military and Social Expenditures.* Available from World Priorities, Box 25140, Washington, DC 20007. A stunning annual report comparing the world's investment in war with its investments in education, health care, and economic development.

FAO. *The State of Food and Agriculture.* Rome: United Nations.

UNEP. *Environmental Data Report.* Oxford: Basil Blackwell, Ltd., published annually.

UNFPA. *The State of World Population.* United Nations Population Fund, 220 East 42nd Street, New York, NY 10017.

UNICEF. *The State of the World's Children.* New York: Oxford University Press.

World Bank. *World Development Report.* New York: Oxford University Press. These annual reports from the United Nations System contain not only world data, but also analyses of how the U.N. agencies view the data and what their primary policy agendas are.

World Resources Institute. *World Resources.* New York: Oxford University Press. The World Resources report appears every second year. It contains comprehensive summaries of the state of population and health, human settlements, food and agriculture, forests and rangeland, wildlife, energy, water, atmosphere, etc., and it also features reviews of special topics, such as climate change or ecosystem restoration.

On the Sustainable Society

We can't begin to list here the full outpouring of excellent publications on the philosophy, politics, economics, demographics, energy, and agriculture of a sustainable world. We list here only a few works, some of the most important and some of the most recent. We encourage the reader to plunge into this lively global discussion at any point. These books will lead to you others.

Benedick, Richard Eliot. *Ozone Diplomacy: New Directions in Safeguarding the Planet.* Cambridge: Harvard University Press, 1991. The story of the ozone layer from the first scientific papers to the London agreement, written by one of the major negotiators for the United States. An important piece of history to understand, as preparation for further global agreements to manage problems of pollution on a global scale.

Berry, Thomas. *The Dream of the Earth.* San Francisco: Sierra Club Books, 1988. Thomas Berry is a visionary and theologian who speaks and writes about "a viable mode of human presence upon the earth." This is a collection of noble and inspiring essays, which lays forth a new "story" of the human role in the universe–one that is consistent with ancient religions, with modern science, and with a sustainable society.

Berry, Wendell. *Home Economics.* San Francisco: North Point Press, 1987. Wendell Berry is a poet, novelist, and farmer. He favors especially the "act locally" part of the environmental dictum "think globally, act locally." His writings are all incisive, critical, based on a celebration of good stewardship and good human communities. This is a fine book to start with, but anything by Wendell Berry is a contribution to the thinking that will bring forth a sustainable and satisfying world.

Clark, Mary E. *Ariadne's Thread: The Search for New Modes of Thinking.* New York: St. Martin's Press, 1989. A comprehensive and courageous book that arose out of a course on the global future taught by faculty from fifteen different disciplines at San Diego State University. The book explores new thinking toward sustainability not only in energy and ecology and economics, but also in psychology, anthropology, religion, and government.

Daly, Herman. *Steady-State Economics.* Washington, DC: Island Press, 1991. A collection of essays by the foremost theoretician of the economics of the sustainable society. Easy to read, written for the most part for the general public, and very thought-provoking.

Daly, Herman, and John Cobb. *For the Common Good.* Boston: Beacon Press, 1989. A more technical and thorough book than the one listed above, written for people familiar with the language of professional economics. It analyzes in depth the reasons why current economic theory does not deal with the whole-system requirements of either the human society or the environment, and, without discarding the achievements of modern economics, it begins to lay the groundwork for the additions and corrections that will establish the more complete economics of community and sustainability.

Ehrlich, Paul R., and Anne H. Ehrlich. *Healing the Planet.* Reading, MA: Addison-Wesley, 1991. This husband-wife team has been documenting global problems and pointing toward solutions for decades. Their latest work is a readable and comprehensive summary of how to read the danger signals coming from the planet and what kinds of actions can restore the planetary system and ensure the human future.

Gever, John, Robert Kaufmann, David Skole, and Charles Vorosmarty. *Beyond Oil: The Threat to Food and Fuel in the Coming Decades.* Niwot: University Press of Colorado, 1991. A fascinating analysis of the degree to which the United States is approaching or has exceeded its carrying capacity, especially with respect to energy and agriculture. The study is based upon painstaking use of data and on computer modeling.

IUCN, UNEP, WWF. *Caring for the Earth: A Strategy for Sustainable Living.* London: Earthscan Publications, 1991. A compendium of principles and actions for sustainable living, based on the World Conservation Strategy published in 1980 by the same three international organizations. An excellent summary of the state of global thinking on the nature of the sustainable society and the steps to bring it about.

Lovelock, J. E. *Gaia: A New Look at Life on Earth.* Oxford: Oxford University Press, 1979. A controversial and thought-provoking hypothesis about the earth as a whole system. It has spawned on the one hand a field of serious scientific inquiry into earth systems and on the other hand a new religion worshipping the earth as a being, and a goddess.

Lovins, Amory B. *Soft Energy Paths.* Cambridge: Ballinger, 1977. The book that introduced the central concepts of an energy system for a sustainable world. The numbers in this book are now out of date–improving technology has made the soft energy path much more feasible and economically favorable than it looked in 1977. But the arguments are still valid, and the book is still a classic.

Mathews, Jessica Tuchman, ed., *Preserving the Global Environment: The Challenge of Shared Leadership.* New York: W. W. Norton, 1991. Essays about policy by some of the world's leading thinkers in the fields of population, deforestation, energy, economics, regulation, and international cooperation.

Meadows, Dennis L., ed., *Alternatives to Growth–I.* Cambridge: Ballinger, 1977. Distributed by Heronbrook Publications, P.O. Box 844, Durham, NH 03824. In this book seventeen chapters describe the ways society could exist in harmony with the planet's limits to growth. Four sections of the book focus on "Nutrition and Energy in the Steady State," "Economic Alternatives in an Age of Limits," "The Politics of Equity and Social Progress in a Finite World,' and " Life-styles and Social Norms for a Sustainable State." The book presents the winning papers that were submitted to the George and Cynthia Mitchell International Competition on sustainable futures.

Meadows, Donella H. *The Global Citizen.* Washington, DC: Island Press, 1991. A compilation of newspaper columns written from 1985 to 1990 on the issues raised by growth, limits, and sustainability. Subjects range from personal lifestyle to global policy, from specific issues of energy, agriculture, waste management, and pollution control to concerns about leadership, ethics, and vision. There is a special attempt to include "good news" case studies.

Milbrath, Lester W. *Envisioning a Sustainable Society.* Albany: State University of New York Press, 1989. A detailed description of the sustainable society, as far as anyone can see it now, but with emphasis on the fact that it cannot be fully foreseen and certainly not dictated–it will require, above all, *learning.*

Orr, David W. *Ecological Literacy.* Albany: State University of New York Press, 1992. This collection of penetrating essays on the kind of education needed for sustainability also contains a much more complete "sustainability reading list" then we can include here

Sagoff, Mark. *The Economy of the Earth.* Cambridge: Cambridge University Press, 1988. A philosopher's critique of welfare economics. Thoughts on ethics, politics, law, and economics as if the earth mattered.

Schumacher, E. F. *Small is Beautiful.* New York: Harper & Row, 1973. A classic work of clear thinking about poverty and development, economics in general and resource economics in particular, laced with a welcome thread of philosophical detachment and moral commitment.

Swimme, Brian. *The Universe Is a Green Dragon.* Santa Fe: Bear and Company, 1984. A beautiful interpretation of Thomas Berry's teachings; a new story and new vision of humanity's role on the earth and in the universe.

Wilson, E. O., ed., *Biodiversity.* Washington, DC: National Academy Press, 1988. A collection of papers written by many of the world's experts on biodiversity for a National Forum on Biodiversity, sponsored by the National

Academy of Sciences and the Smithsonian Institution. If you want to know what ecologists are seeing as they monitor the ecosystems of the world and how they feel about it, this is an excellent volume to read.

Woodwell, George, ed., *The Earth in Transition: Patterns and Processes of Biotic Impoverishment.* Cambridge: Cambridge University Press, 1990. Another collection of papers from a congress of field ecologists, held at about the same time as the National Forum described in the entry above. The papers in this volume are on the whole longer and more scholarly than the ones in the book edited by Wilson. Together they add up to a stunning documentation of the erosion of biotic resources in every part of the world, from the tundra to the tropical forest, from the coral reefs to the lakes of Canada, and from the Great Basin of the U.S. West to the eucalyptus forests of Australia.

World Commission on Environment and Development. *Our Common Future.* Oxford: Oxford University Press, 1987. This is the report of the prestigious international panel that conducted a two-year study and held hearings all around the world on the issues of environment and development. The two primary contributions of this study were the definition and popularization of the idea of sustainability and the strong linkage of the issues of environment and development. The book is full of interesting data and also of poignant quotations from people who testified at the hearings.

Some Works by and for The Club of Rome

There have been numerous reports *to* The Club of Rome by research groups commissioned by them, as we were commissioned twenty years ago, to investigate some aspects of what the Club calls the "global *problematique.*" A few books have been written *by* The Club of Rome or by some of its prominent members in the Club's name. Some of the most recent and important examples of the latter we list here.

King, Alexander, and Bertrand Schneider. *The First Global Revolution.* New York: Pantheon Books, 1991. From the *problematique* to the *resolutique.* A "blueprint for the twenty-first century" by the Council of The Club of Rome.

Peccei, Aurelio. *The Human Quality.* Oxford: Pergamon Press, 1977. In the first chapter of this book the founder of The Club of Rome writes a short autobiography, which is a testament to the nobility of the human race, and a gripping story of a man who experienced everything from Fascist torture chambers to the boardrooms of major corporations. The rest of the book is his account of the founding of the Club, the preparation of *The Limits to Growth,* and the human revolution he believed was both necessary and possible.

Peccei, Aurelio. *One Hundred Pages for the Future.* New York: Pergamon Press, 1981. Less personal than *The Human Quality,* more focused on the state of the world and what to do about it. Peccei consistently points out that the solutions to the world's problems are to be found within ourselves.

Peccei, Aurelio, and Daisaku Ikeda. *Before It Is Too Late.* Tokyo: Kodansha International, 1984. A conversation on global issues between the founder of The Club of Rome and a Buddhist lay leader. The last publication of Aurelio Peccei before his death in 1984.

Pestel, Eduard. *Beyond the Limits to Growth.* New York: Universe Books, 1989. Another early member of The Club of Rome, a global modeler himself, provides his memories of the Club's activities and sets forth his view of where to go from here.

Glossary of Systems Terms

behavior The performance of a system over time—growth, steady equilibrium, oscillation, decline, randomness, evolution, chaos, or any complex combination of these behaviors.

collapse An uncontrolled decline in a population or economy induced when that population or economy overshoots the sustainable limits to its environment and in the process reduces or erodes those limits. Collapse is especially likely to occur when there are positive loops of erosion, so that a degradation of the environment sets in motion processes that degrade it further.

delay A time lag between a cause and an effect. It can occur because of a time-consuming intervening physical process. For example there is a construction delay between the initial investment in an electric power plant and its completion, and a delay between the application of a pesticide to the soil and its eventual percolation into the groundwater. Delays can also occur in information flows—for example, the normal "noise" or variation in the weather means that weather patterns must be averaged over many years before there is reliable information about whether a climate change has occurred.

equilibrium When a stock's inflows equal its outflows (see **stock**). A population is in equilibrium when its births plus inmigrations equal its deaths plus outmigrations. A lake is in equilibrium when its inflows equal its outflows plus evaporation. In equilibrium, the contents of the stock are continually changing, but its overall level is constant.

erosion A decline in the resource base supporting a system that in itself can lead to further decline. A positive feedback loop going downward, so that each decrease makes the next decrease likely.

exponential growth Growth by a constant fraction of the growing quantity during a constant time period. Money in the bank grows exponentially when interest is added at the rate, say, of 7% of whatever is already in bank every year. Populations grow exponentially when they multiply by a fraction of themselves every year, or every month, or, in the case of microbes, every few minutes. When something grows exponentially, it continuously doubles—2, 4, 8, 16, 32—with a characteristic doubling time.

feedback loop A closed chain of causal connections. Generally feedback loops proceed from a stock, through a set of decision or actions dependent on the condition of that stock, and back again to a change in the stock.

flow A rate of change of a stock, usually an actual physical flow into or out of a stock. Whatever units a stock is measured in, all flows into and out of that stock are measured in the same units per unit of time. Important flows in World3 are human births per year and deaths per year, capital investment per year and depreciation per year, pollution generated per year and absorbed per year, and nonrenewable resources consumed per year.

negative feedback loop A chain of cause-and-effect relationships that propagates a change in one element around a circle of causation until it comes back to change that element in a direction opposite to the initial change. Whereas positive loops generate runaway growth, negative feedback loops tend to regulate growth, to hold a system within some acceptable range, or to return it to a stable state.

nonlinearity A relationship between a cause and an effect that is not linear, which is to say, not strictly proportional for all values of the cause or the effect. For instance, suppose you put 2 pounds of fertilizer on your garden and your yield went up 10%, and then you put 4 pounds of fertilizer on and your yield went up 20%. So far the relationship between fertilizer and yield is linear. But it is quite unlikely that applying 200 pounds of fertilizer would raise your yield 1000% (it could kill your garden entirely!). Over that range the relationship between fertilizer and yield is nonlinear.

overshoot To go beyond a target, and in the particular meaning of this book, it is to go beyond the sustainable carrying capacity of the environment. Overshoot is caused by delays or faults in feedback information that do not allow a system to control itself relative to its limits. Overshoot is also a function of the speed of change or movement of the system—a feedback delay that can be accommodated at low speed may cause overshoot at higher speed.

positive feedback loop A chain of cause-and-effect relationships that closes in on itself so that an increase in any one element in the loop will start a sequence of changes that will increase the original element even more. A positive feedback loop can be a "vicious circle," or a "virtuous circle," depending on whether the growth it produces is wanted or not.

sink* The ultimate destination of material or energy flows used by a system. The atmosphere is the sink for carbon dioxide generated by burning coal. A municipal landfill is often the sink for paper made from wood from a forest.

source* A point of origin of material or energy flows used by a system. Coal deposits under the ground are the sources of coal in the short term; in the very long term forests are the sources of coal. Forests are sources of wood in the short term; in the intermediate term soil nutrients, water, and solar energy are the sources of forests.

stock An accumulation, store, level, or quantity of material, energy, or information. It represents the current state of the system; it reflects the history of flows into and out of that stock; and, since stocks typically change only slowly over time, it can act as a delay in the system's response. Important stocks in World3 are population, industrial capital, service capital, agricultural land, pollution, and nonrenewable resources.

structure The entire set of stocks, flows, feedback loops, and delays that define all the interconnections of a system. A system's structure determines its full range of behavioral possibilities. The actual behavior at any given moment arises from the system's structure plus its environment plus its current internal state.

system An interconnected set of elements that is coherently organized around some purpose. A system is more than the sum of its parts. It

*When you look carefully at sources and sinks, and especially when you look at them over the long term, you see that they are not things, like buckets that can be filled or emptied, but processes. They are buckets that are being continually refilled or emptied by nature at varying rates. Sources and sinks are limits to systems, but they are ultimately limits to the rates at which things can happen, not to the amount that can happen.

can exhibit dynamic, adaptive, goal-seeking, self-preserving, and evolutionary behavior.

throughput The flow of energy and/or material from the original sources, through a system (where it may be transformed), and out to the ultimate sinks.

List of Tables and Figures with Sources

Chapter 1

Motor Vehicles Manufacturers Association of the U.S., *World Motor Vehicle Data 1990* (Washington, DC); Motor Vehicles Manufacturers Association of the U.S., *World Motor Vehicle Data 1976* (Washington, DC); Energy Information Administration, U.S. Dept. of Energy, *Annual Energy Review 1989* (Washington, DC: Government Printing Office); Energy Information Administration, U.S. Dept. of Energy, *Monthly Energy Review* (Washington, DC: Government Printing Office, May 1991); Oil & Gas Journal, *Energy Statistics Sourcebook 1990* (Tulsa, OK: PennWell Publishing Co., 1991); U.S. Bureau of the Census, *Statistical Abstracts of the United States 1990* (Washington, DC: Government Printing Office); American Iron & Steel Institute, *Annual Statistical Report 1990* (New York); U.S. Bureau of Mines, *Mineral Commodity Summaries 1990* (Washington, DC: Government Printing Office); U.S. Bureau of Mines, *Mineral Facts & Problems 1970* (Washington, DC: Government Printing Office); *Yearbook of World Energy Statistics 1979* (New York: United Nations, 1981); *Yearbook of World Energy Statistics 1988* (New York: United Nations); *Energy Statistics Yearbook 1989* (New York: United Nations,

Chapter 2

Chapter 2

Food and Agriculture 1991, Proceedings of the 26th Session (Rome: United Nations, November 1991), 9–28.

Figure 2-2 16
WORLD URBAN POPULATION
World Population Prospects 1990 (New York: United Nations); Population Reference Bureau, *1991 World Population Data Sheet* (Washington, DC, 1991).

Figure 2-3 17
LINEAR VERSUS EXPONENTIAL GROWTH OF SAVINGS

Figure 2-4 25
WORLD DEMOGRAPHIC TRANSITION
The World Population Situation 1970, Population Studies no. 49 (New York: United Nations, 1971); *World Population Prospects 1990* (New York: United Nations).

Figure 2-5 26
AVERAGE ANNUAL POPULATION INCREASE
The Determinants and Consequences of Population Trends, Population Studies no. 50 (New York: United Nations, 1973); *World Population Prospects 1990* (New York: United Nations); Edward Bos et al., *Asia Region Population Projections, 1990–91,* Working Papers series 599 (Population, Health, and Nutrition Division, Population and Human Resources Dept. of Policy, Research, and External Affairs, The World Bank, February 1991).

Figure 2-6a 30
DEMOGRAPHIC TRANSITIONS IN INDUSTRIALIZED NATIONS
Figure 2-6b 31
DEMOGRAPHIC TRANSITIONS IN LESS-INDUSTRIALIZED NATIONS
Demographic Yearbook 1950 and subsequent years (New York: United Nations); R. A. Easterlin, ed., *Population and Economic Changes in Developing Countries* (Chicago: University of Chicago Press, 1980); J. Chesnais, *La Transition Demographique* (Paris: University of France Press, 1986); Nathan Keyfitz and W. Flieger, *World Population: An Analysis of Vital Data* (Chicago: University of Chicago Press, 1968); Population Reference Bureau, *1991 World Population Data Sheet* (Washington, DC, 1991); U.K. Office of Population Censuses & Surveys, *Population Trends,* no. 52 (London: H.M.S.O., June 1988).

Figure 2-7 32
BIRTH RATES AND GNP PER CAPITA IN 1989
Population Reference Bureau, *1991 World Population Data Sheet* (Washington, DC, 1991); Population Reference Bureau, *1989 World Population Data Sheet* (Washington, DC, 1989); CIA, *Handbook of Economic Statistics 1990* (Washington, DC, September 1990).

Chapter 3

Table 3-4 102
THE ENVIRONMENTAL IMPACT OF POPULATION, AFFLUENCE, AND TECHNOLOGY
Amory Lovins, Rocky Mountain Institute, Snowmass, CO, 1990.

Figure 3-1 45
POPULATION AND CAPITAL IN THE GLOBAL ECOSYSTEM
R. Goodland, H. Daly, and S. El Serafy, "Environmentally Sustainable Economic Development Building on Bruntland," Environment Working Paper no. 46 (The World Bank: July 1991).

Figure 3-2 48
WORLD GRAIN PRODUCTION 1950–1990
FAO, *Production Yearbook 1951* and subsequent years (Rome: United Nations); FAO, Quarterly Bulletin of Statistics 4, no. 2 (Rome: United Nations, 1991).

Figure 3-3 49
GRAIN YIELDS
FAO, *Production Yearbook 1970* and subsequent years (Rome: United Nations).

Figure 3-4 51
POSSIBLE LAND FUTURES
Land estimates from: G. M. Higgins et al., *Potential Population Supporting Capacities of Lands in the Developing World* (Rome: FAO, 1982); World Resources Institute, *World Resources 1990-91* (New York: Oxford University Press, 1990); Population estimates from: Rodolfo A. Bulatao et al., *World Population Projections 1989–90,* The World Bank (Baltimore: Johns Hopkins University Press, 1990).

Figure 3-5 55
FRESH WATER RESOURCES
Robert P. Ambroggi, "Water," *Scientific American* (September 1980): 103.

Figure 3-6 59
FOREST COVER IN COSTA RICA 1940–1984
Carlos Quesada, ed., Estrategia Conservaeión para el Desarrollo Sostenible en Costa Rica, ECODES (San José: Ministerio de Recursos Naturales, Energía y Minas).

Figure 3-7 61
SOME POSSIBLE PATHS OF TROPICAL DEFORESTATION

Figure 3-8 62
WORLD ROUNDWOOD PRODUCTION
FAO, *Forest Production Yearbook 1960* and subsequent years (Rome: United

Chapter 4

Chapter 5

Table 5-1 143

USES, PRODUCTION RATES, AND RESIDENCE TIMES OF THE IMPORTANT OZONE-DEPLETING CHEMICALS

A. Makhijani, A. Makhijani, and A. Bickel, *Saving Our Skins: Technical Potential and Policies for the Elimination of Ozone-depleting Chlorine Compounds* (Environmental Policy Institute and the Institute for Energy and Environmental Research, September 1988); Environmental Studies Board of the National Research Council, *Causes and Effects of Changes in Stratospheric Ozone: 1983 Update* (Washington, DC: National Academy Press, 1984).

Figure 5-1 144

WORLD REPORTED PRODUCTION OF CFC-011 AND CFC-012

Chemical Manufacturers Association, *1989 Production and Sales of Chlorofluorocarbons 11 & 12* (CMA Fluorocarbon Program Panel, December 1990); Mack McFarland, "Chlorofluorocarbons and Ozone: 1st Plenary Jekyll Is. Meeting," *Environmental Science and Technology* 23, no. 10 (1989).

Figure 5-2 146

ABSORPTION OF ULTRAVIOLET LIGHT BY THE ATMOSPHERE

UNEP, *The Ozone Layer*, UNEP/GEMS Library no. 2 (Nairobi, 1987).

Figure 5-3 149

HOW CFCS DESTROY STRATOSPHERIC OZONE

Figure 5-4 151

OZONE MEASUREMENTS AT HALLEY BAY, ANTARCTICA

J. C. Farman, B. G. Gardiner, and J. D. Shanklin, "Large Losses of Total Ozone in Antarctica Reveal Seasonal ClO_x/NO_x Interaction," *Nature* 314 (1985): 207–210.

Figure 5-5 152

AS REACTIVE CHLORINE INCREASES, ANTARCTIC OZONE DECREASES

J. G. Anderson, W. H. Brune, and M. H. Proffitt, "Ozone Destruction by Chlorine Radicals within the Antarctic Vortex: The Spatial and Temporal Evolution of $ClO-O_3$ Anticorrelation Based on in Situ ER-2 Data," *Journal of Geophysical Research*, 94, no. D9 (30 August 1989): 11,465–11,479.

Figure 5-6 156

REAL AND PROJECTED GROWTH OF STRATOSPHERIC INORGANIC CHLORINE CONCENTRATIONS DUE TO CFC EMISSIONS

John S. Hoffman and Michael J. Gibbs, *Future Concentrations of Stratospheric Chlorine and Bromine*, Office of Air and Radiation, U.S. Environmental Protection Agency 400/1-88/005 (Washington, DC: Government Printing Of-

Chapter 6

Chapter 7

Index

E

F

E

F

M

N

T

U

V

W